PRINCIPLES AND PRACTICE

Springer

Berlin
Heidelberg
New York
Barcelona
Hong Kong
London
Milan
Paris
Singapore
Tokyo

Peter James

Proteome Research: Mass Spectrometry

With 77 Figures

 Springer

Dr. PETER JAMES
ETH-Zentrum
Protein Chemistry Laboratory
Universitätsstraße 16
8092 Zürich
Switzerland

QP
551
.P7558
2001

ISBN 3-540-67255-9 Springer-Verlag Berlin Heidelberg New York

Library of Congress Cataloging-in-Publication Data

Proteome research : mass spectrometry / P. James [editor].
  p. cm. -- (Principles and practice)
  Includes bibliographical references and index.
  ISBN 3540672559 (hardcover : acid-free paper) -- ISBN 3540672567 (softcover :
acid-free paper)
  1. Proteins--Analysis. 2. Mass spectrometry. I. James, P. (Peter), 1961- II. Series.

QP551.P7558 2000.
572'.636--dc21

Springer-Verlag Berlin Heidelberg New York
a member of BertelsmannSpringer Science+Business Media GmbH

© Springer-Verlag Berlin Heidelberg 2001
Printed in Germany

The use of general descriptive names, registered names, trademarks, etc. in this publication does not imply, even in the absence of a specific statement, that such names are exempt from the relevant protective laws and regulations and therefore free for general use.

Cover design: D&P, Heidelberg
Typesetting: Best-set Typesetter Ltd., Hong Kong
SPIN 10724397   39/3130 – 5 4 3 2 1 0 – Printed on acid-free paper

*To Evelyne and Nina
for their patience and love.*

Acknowledgements

This book marks the end of an era, that of the A-floor folk. It was both a scientifically highly productive and intellectually stimulating 8 years. More importantly, it was fun, and the atmosphere was unique. Many people would do well to learn that good science does not spring from a sterile environment and that flexibility breeds trust, which is the greatest form of motivation. A large amount of humour and the feeling that we were working with friends rather than colleagues made this time unforgettable, and it is fitting that we finish it with a book such as this.

Contents

1 **Mass Spectrometry and the Proteome**
PETER JAMES

1.1 The Industrialisation of Biology 1
1.2 The Proteome 2
1.2.1 mRNA and Protein 2
1.2.2 Visualising the Proteome: Two-Dimensional Gel
Electrophoresis 2
1.3 Mass Spectrometry and Proteomics 3
1.3.1 General Requirements 3
1.3.2 Sensitivity 3
1.3.3 Sample Handling 5
1.3.4 High-Throughput Protein Identification 5
1.3.5 Quantification 5
1.3.6 Post-Translational Modifications 6
1.4 Aims of the Book 6
References 8

2 **Basics of Triple-Stage Quadrupole/Ion-Trap Mass
Spectrometry: Precursor, Product and Neutral-Loss Scanning.
Electrospray Ionisation and Nanospray Ionisation**
DAVID ARNOTT

2.1 Introduction 11
2.2 Fundamental Principles 12
2.2.1 Mass Analysis 12
2.2.2 Collision-Induced Dissociation 15
2.2.3 Electrospray Ionisation 16
2.3 Triple-Quadrupole MS 17
2.3.1 Instrument Overview 17
2.3.2 Product-Ion Scans 18
2.3.3 Precursor-Ion Scans 21
2.3.4 Neutral-Loss Scans 23
2.4 Ion-Trap MS 24
2.4.1 Instrument Overview 24
2.4.2 Full Mass-Range and High-Resolution Scans 25
2.4.3 Product-Ion Scans 26

2.4.4   MS^n 27
2.5   Virtues of TSQ and Ion-Trap Instruments 28
         References 29

**3   The Basics of Matrix-Assisted Laser Desorption,
         Ionisation Time-of-Flight Mass Spectrometry and Post-Source
         Decay Analysis**
         BERNHARD SPENGLER

3.1   Introduction 33
3.2   Principles of MALDI 35
3.2.1   Ion Formation by MALDI 36
3.2.2   Sample Preparation 39
3.2.3   Instrumentation 40
3.2.3.1 Ion Source 40
3.2.3.2 Mass Analyser 40
3.2.3.3 Ion Detector 41
3.2.3.4 Delayed Ion Extraction 42
3.3   Principles of MALDI-PSD Mass Analysis 43
3.3.1   Mechanisms for Ion Activation 43
3.3.2   Instrumentation for MALDI-PSD 46
3.3.2.1 Ion Source 46
3.3.2.2 Ion Gate 47
3.3.2.3 Mass Analyser 47
3.3.3   Interpretation of PSD Mass Spectra 48
3.3.4   Improving Spectral Information 50
3.4   Conclusion 50
         References 51

**4   Data-Controlled Micro-Scale Liquid Chromatography –
         Tandem Mass Spectrometry of Peptides and Proteins:
         Strategies for Improved Sensitivity, Efficiency and Effectiveness**
         DOUGLAS C. STAHL and TERRY D. LEE

4.1   Introduction 55
4.2   The ES Process 56
4.3   Micro-ES Emitter Design 57
4.3.1   Emitter Construction 57
4.4   High-Voltage Connection 58
4.5   Capillary Columns 60
4.5.1   Column Construction 60
4.6   Gradient-Delivery Systems 62
4.6.1   Flow Splitting 62
4.7   Sample Loading 63
4.8   Variable-Flow Chromatography and Peak Parking 65
4.8.1   Variable-Flow Chromatography 65

4.8.2	Peak Parking	67
4.8.3	Automation	68
4.9	Data-Controlled Analysis	69
4.9.1	Analysing Peptide Modifications	70
4.9.2	Software Integration	71
4.10	Conclusion	72
	Acknowledgements	72
	References	73

5 Solid-Phase Extraction–Capillary Zone Electrophoresis–Mass Spectrometry Analysis of Low-Abundance Proteins

DANIEL FIGEYS and RUEDI AEBERSOLD

5.1	Introduction	75
5.2	Fabrication of the SPE-CZE System	76
5.2.1	General Design	76
5.3	Procedure for the SPE-CZE-MS/MS Experiment	79
5.3.1	Pre-Saturation	79
5.3.2	Sample Loading	79
5.3.3	Sample Elution and Separation	79
5.3.3.1	Solid-Phase Extraction–Capillary Zone Electrophoresis	79
5.3.3.2	SPE and Transient Isotachophoresis	81
5.3.4	Sample Overloading	81
5.3.5	MS and Data Processing	82
5.4	System Optimisation and Variations of the Basic Method	84
5.4.1	Extraction Materials for the SPE Cartridge	84
5.4.2	Surface Chemistry for Capillary Tubing	87
5.4.3	Separation	88
5.4.4	Peak-Parking Mode	92
5.4.4.1	Instrument-Control Procedures	93
5.4.4.2	Analysis of Protein with the Peak-Parking Mode	94
5.5	Conclusion	94
	Appendix: Protocols	96
1	SPE-CZE (Quick Guide)	96
2	SPE-CZE-Transient Isotachophoresis (Quick Guide)	96
3	Coating of Capillaries: General Comments	96
4	γ-Methacrylopropyl Trimethoxysilane Coating	97
5	MAPTAC Coating with Si–O–Si Attachments	97
6	MAPTAC Coating with Si–C Attachments	98
	References	98

6 Protein Identification by Peptide-Mass Fingerprinting

PAOLA DAINESE and PETER JAMES

6.1	Introduction	103
6.1.1	Protein Identification Using Proteolytic Masses	103

6.2	Database Searching and Scoring Schemes	105
6.3	Strengths and Limitations of the Method	106
6.3.1	Mass Accuracy and Automation	107
6.3.2	Multiple Proteins in a Sample	108
6.3.3	Protein Modifications	109
6.3.4	Removing Post-Translational Modifications for Searching	110
6.3.5	Relative Effectiveness of Different Digesters	110
6.3.6	Practical Limitations of Digestion	111
6.4	Extensions of the Method	112
6.4.1	Orthogonal Data	112
6.4.2	Cross-Species Identification	113
6.4.3	Chemical Modifications	114
6.4.4	Iterative Searching	114
6.5	Sequence Tagging	116
6.5.1	Ragged Termini	116
6.5.2	Enzymatic-Ladder Sequencing	117
6.5.3	Chemical-Ladder Sequencing	117
6.6	Conclusion	119
	References	121

7   Protein Identification by SEQUEST
DAVID L. TABB, JIMMY K. ENG, and JOHN R. YATES III

7.1	Introduction to SEQUEST and Protein Identification	125
7.1.1	Peptide MS and the Collision-Induced Dissociation Process	126
7.1.2	Interpretation of Tandem Mass Spectra	126
7.2	Description of the Algorithm	128
7.2.1	Spectrum Pre-Processing	128
7.2.2	Selection of Candidate Sequences from a Database	128
7.2.3	Preparation for Spectrum Comparison	129
7.2.4	Cross-Correlation	129
7.3	Interpretation of SEQUEST Scores	130
7.3.1	Interpretation of a Spectrum's SEQUEST Results	130
7.3.2	Interpretation of an Entire LC/MS/MS Analysis	132
7.3.3	Interpretation of Multiple LC/MS/MS Analyses	132
7.4	Extensions of Functionality	135
7.4.1	Comprehensive Modification Analysis	135
7.4.2	Performance Improvements	136
7.4.2.1	Contaminant Filtering	136
7.4.2.2	Spectral-Quality Filtering	137
7.4.2.3	High-Performance Computing Using a Parallel Cluster	138
7.5	Applications of SEQUEST	140
7.5.1	Complex-Mixture Analysis	140
7.5.2	Subtractive Analysis	140

7.5.3   Case Studies from in-Gel Digest Analysis 141
7.6   Conclusion 142
          References 142

8   **Interpreting Peptide Tandem Mass-Spectrometry**
          Fragmentation Spectra
          WERNER STAUDENMANN and PETER JAMES

8.1   Introduction to Peptide Tandem Mass Spectrometry 143
8.1.1   Peptide Ionisation 143
8.1.2   Peptide Fragmentation 144
8.1.3   Mechanism of Amide-Bond Cleavage 146
8.1.4   Sequence-Dependant Fragmentation Types 148
8.1.4.1   Immonium Ions 148
8.1.4.2   Internal Cleavage 148
8.1.4.3   Preferential Cleavage C-Terminal to Asp 149
8.1.4.4   Low-Energy Production of a-Type Ions 150
8.1.4.5   Loss of C-Terminal Amino Acids 150
8.1.4.6   Amino Acid-Specific Mass Losses 150
8.2   Manual Interpretation of Peptide MS/MS Spectra 152
8.2.1   General Considerations 152
8.2.2   An Approach to Manual Interpretation 153
8.2.2.1   Interpretation Checklist 153
8.2.3   Interpreting a Simple MS/MS Spectrum
          of a Tryptic Peptide 155
8.2.4   The Importance of Charge Location 157
8.3   Aids to Spectral Interpretation and the Avoidance
          of Pitfalls 159
8.3.1   Distinguishing Between Residues with Closely
          Related Masses 159
8.3.2   Isotopic Labelling of Ion Series 160
8.3.3   A Cautionary Tale 161
8.4   Manual Interpretation and Database Searching 161
8.5   Conclusion 164
          Acknowledgements 164
          References 165

9   **Automated Interpretation of Peptide Tandem Mass Spectra**
          and Homology Searching
          RICHARD S. JOHNSON

9.1   Database-Searching Programs for Inexact Peptide Matches ... 167
9.2   Algorithms for Automated Interpretation of Peptide
          Tandem Mass Spectra 168
9.2.1   Brute-Force Method 168
9.2.2   Sub-Sequencing Approach 168

9.2.3   Graphical Display to Assist Manual Interpretation 169
9.2.4   Graph-Theory Approach 169
9.3   Ambiguities Associated with Sequence Determinations
         from MS/MS Spectra 172
9.3.1   Isomeric and Isobaric Amino Acid Residues 172
9.3.2   Amino Acid Residues That Are Isobaric
         with Dipeptide Residues 173
9.3.3   Missing Fragmentation 173
9.3.4   Unknown and Unusual Fragmentation Reactions
         and Artefacts 173
9.3.5   Inherent Ambiguity Due to Spectrum Complexity 174
9.4   Using Ambiguous Peptide Sequences Derived
         from Tandem Mass Spectra 174
9.4.1   Automated Validation of Database Matches 174
9.4.2   Use of a Homology-Based Search Program That Has
         Been Modified to Handle Sequence Ambiguities 176
9.4.2.1 Identifying Peptides Derived from Non-Consensus Cleavages
         or from Chemical Artefacts 176
9.4.2.2 Database Sequence Errors and Homologous Proteins 178
9.5   Resolving Sequence Ambiguities 178
9.5.1   Higher Mass-Accuracy Measurements of Fragment Ions
         and Peptide Molecular Weights 179
9.5.2   Deuterium Incorporation 179
9.5.3   Chemical Modifications 180
9.5.4   C-Terminal ^{18}O Labelling 180
9.5.5   Multiple Stages of MS/MS 180
9.5.6   Comparison with Synthetic Peptides 182
9.6   Conclusion 182
         Acknowledgements 182
         References 183

**10   Specific Detection and Analysis of Phosphorylated Peptides
by Mass Spectrometry**
MANFREDO QUADRONI

10.1   Introduction 187
10.1.1   Experimental Approaches to the Study of Phosphorylation
         Sites in Proteins 188
10.2   Specific Detection of Phosphopeptides in Complex
         Peptide Mixtures 189
10.2.1   Isolation of Phosphopeptides Prior to MS Analysis 189
10.2.2   Enzymatic Removal of the Phosphate Group 190
10.2.3   Phosphopeptide Detection by Selective MS
         Scanning Methods 190
10.2.4   Phosphopeptide Detection by Selective MS Scanning
         Methods After Specific Chemical Modifications 192

10.3	Phosphorylation-Site Analysis by MS/MS	196
10.3.1	Phosphorylation-Site Analysis of Native Phosphopeptides ...	196
10.3.2	Phosphorylation-Site Analysis after Chemical Modification of the Phosphoamino-Acid Side Chains	197
10.4	Data Analysis	199
10.5	Conclusion	202
	Appendix	203
1	IMAC Chromatography for the Isolation of Phosphorylated Peptides	203
1.1	Column Packing	203
1.2	IMAC Separation	203
2	Chemical Modification of Phosphopeptides for MS	204
2.1	Synthesis of 2-[4-pyridyl]ethanethiol	204
2.2	Conversion of Phosphoserine to S-Pyrydylethylcysteine	204
	References	204

11   Glycoproteomics: High-Throughput Sequencing of Oligosaccharide Modifications to Proteins
PAULINE M. RUDD, CRISTINA COLOMINAS, LOUISE ROYLE, NEIL MURPHY, EDMUND HART, ANTHONY H. MERRY, HOLGER F. HEBERSTREIT, and RAYMOND A. DWEK

11.1	Beyond the Genome and the Proteome Lies the Glycome	207
11.1.1	The Importance of Glycosylation in Proteomics	207
11.1.2	High-Throughput Analysis of Heterogeneous Glycans Requires New Analytical Strategies	208
11.2	A Rapid and Robust Strategy for N- and O-Glycan Analysis	209
11.2.1	Release of N- and O-Linked Glycans from Glycoproteins	209
11.2.1.1	Chemical Release of N- and O-Glycans from Glycoproteins Using Anhydrous Hydrazine	209
11.2.1.2	Enzymatic Release and Analysis of N-Linked Oligosaccharides from Protein Bands on Sodium Dodecyl Sulphate Polyacrylamide-Gel Electrophoresis Gels	209
11.2.1.3	Advantages of the "In-Gel" Release Method	212
11.2.2	Labelling the Glycan Pool	213
11.2.3	Resolving Released Glycan Pools and Assigning Structure by HPLC	214
11.2.3.1	Weak Anion-Exchange Chromatography	214
11.2.3.2	Normal-Phase HPLC	215
11.2.3.3	Reverse-Phase HPLC	216
11.2.4	Simultaneous Sequencing of Oligosaccharides Using Enzyme Arrays	216
11.2.5	Analysis of the NP-HPLC Results	219
11.3	Applications of This Technology	221
11.3.1	IgG Glycosylation and Disease	221

11.3.2   Potential Roles for the Glycans Attached to CD59 222
11.3.3   The Three-Dimensional Structure of the Protein Influences
         Glycan Processing 225
11.3.4   Simultaneous Analysis of the N-Glycan Pool from Rat CD48
         Expressed in CHO Cells Reveals Extensive Heterogeneity 225
11.3.5   Analysis of the O-Glycans from Human Neutrophil Gelatinase
         B Suggests That They May Produce an Extended and Rigid
         Region of the Peptide 226
11.4   Conclusion 226
         References 226

12   Proteomics Databases

HANNO LANGEN and PETER BERNDT

12.1   Introduction 229
12.2   Protein- and Nucleotide-Sequence Databases 230
12.2.1   Swiss-Prot 230
12.2.2   TrEMBL 232
12.2.3   GenBank 232
12.2.4   PIR 233
12.2.5   OWL 233
12.2.6   EST databases 233
12.3   Two-Dimensional Gel Protein Databases 234
12.3.1   Swiss 2D Polyacrylamide-Gel Electrophoresis 234
12.3.2   Danish Centre for Human Genome Research 2D-PAGE
         Databases 239
12.3.3   Argonne Protein-Mapping Group Server 240
12.3.4   SIENA 2D-PAGE 240
12.3.5   HEART 2D-PAGE 241
12.3.6   HSC 2D-PAGE 241
12.3.7   National Institutes for Mental Health/National Cancer Institute
         Protein-Disease Database and the National Cancer Institute/
         Frederick Cancer Research and Development Center
         Laboratory of Mathematical Biology Image Processing 241
12.3.8   Yeast-Proteome Database 242
12.3.9   MitoDat 242
12.3.10 Large-Scale Biology 242
12.3.11 UCSF 2D-PAGE 242
12.3.12 ECO2DBASE 243
12.3.13 Embryonic Stem Cells 243
12.3.14 Human Colon-Carcinoma Protein Database 243
12.3.15 Yeast 2D-PAGE 243
12.3.16 Haemophilus 2DE Protein Database 244
12.4   Proteomics-Database Design 244
12.4.1   The Database Schema 245

12.4.1.1 General Remarks 245
12.4.1.2 Sample and Gel Tracking 245
12.4.1.3 Spot Location and Quantification Data 247
12.4.1.4 Mass-Spectrometric Data 248
12.4.1.5 Protein Identifications 249
12.4.2   Database Interface 249
12.4.2.1 Browser 250
12.4.2.2 Communicators 251
12.4.2.3 Editor 251
12.4.2.4 Agents 251
12.4.3   Database Tasks 252
12.4.3.1 Mapping of Proteomes 252
12.4.3.2 Protein-Expression Analysis 252
12.4.3.3 Integrated Genomic Analysis 252
12.4.3.4 Protein Modifications 253
          References 253

13   Quo Vadis
PETER JAMES

13.1   Introduction 259
13.2   Plug-and-Play MS 260
13.3   Recent Advances in MS Instrumentation 261
13.3.1   Sensitivity and Duty Cycles 261
13.3.2   Miniaturisation 262
13.4   Fourier-Transform Ion-Cyclotron Resonance
          Mass Spectrometry 262
13.4.1   The Advantages of FT-ICR MS 262
13.4.1.1 Extreme Sensitivity and Resolution 262
13.4.1.2 Non-Destructive Measurement 263
13.4.2   The Disadvantages of FT-ICR MS 264
13.4.2.1 High Vacuum 264
13.4.2.2 Duty Cycle 264
13.5   Future Directions 265
13.5.1   Bypassing Proteolysis: Fragmenting Whole Proteins
          in the Mass Spectrometer 265
13.5.1.1 Fragmentation Methods 265
13.5.1.2 Protein Identification 265
13.5.2   An Alternative to 2D-PAGE? MS as the Second Dimension ... 266
13.5.3   Molecular-Interaction Mapping 266
13.6   Conclusion 267
          References 268

          Subject Index 271

List of Contributors

RUEDI AEBERSOLD
Dept. of Molecular Biotechnology, K327, Health Sciences Building,
Box 357730, University of Washington, Seattle, WA 98195-7730,
USA
e-mail: ruedi@u.washington.edu

DAVID ARNOTT
Protein Chemistry Department, Genentech, Inc., 1 DNA Way,
South San Francisco, CA 94080, USA
e-mail: arnott@gene.com

PETER BERNDT
Hoffmann-La Roche, Grenzacherstrasse 183, 4002 Basel, Switzerland
e-mail: peter.berndt@Roche.com

CRISTINA COLOMINAS
The Glycobiology Institute, Department of Biochemistry, South Parks Road,
Oxford OX1 3QU, UK
e-mail: cristina@glycob.ox.ac.uk

PAOLA DAINESE
Protein Chemistry Laboratory, Universitaetstrasse 16, ETH-Zentrum,
8092 Zurich, Switzerland
e-mail: paola.dainese@bc.biol.ethz.ch

RAYMOND A. DWEK
The Glycobiology Institute, Department of Biochemistry, South Parks Road,
Oxford OX1 3QU, UK
e-mail: raymond.dwek@exeter.ox.ac.uk

JIMMY K. ENG
University of Washington, Department of Molecular Biotechnology,
Box 357730, Seattle, WA 98195-7730, USA
e-mail: engj@u.washington.edu

DANIEL FIGEYS
National Research Council Canada, Institute for Marine Biosciences,
1411 Oxford Street, Halifax, Nova Scotia, B3H 3Z1, Canada
e-mail: Daniel.figeys@NRC.CA

EDMUND HART
The Glycobiology Institute, Department of Biochemistry, South Parks Road,
Oxford OX1 3QU, UK
e-mail: edhart@bioch.ox.ac.uk

HOLGER F. HEBERSTREIT
The Glycobiology Institute, Department of Biochemistry, South Parks Road,
Oxford OX1 3QU, UK
e-mail: holger@glycob.ox.ac.uk

PETER JAMES
ETH-Zentrum, Protein Chemistry Laboratory, Universitätstrasse 16,
8092 Zürich, Switzerland
e-mail: peter.james@bc.biol.ethz.ch

RICHARD S. JOHNSON
Immunex Corporation, 51 University Street, Seattle, WA 98101-2936, USA
e-mail: rjohnson@immunex.com

HANNO LANGEN
Hoffmann-La Roche, Grenzacherstrasse 183, 4002 Basel, Switzerland
e-mail: hanno.langen@Roche.COM

TERRY D. LEE
Division of Immunology, Beckman Research Institute of the City of Hope,
1450 East Duarte Road, Duarte, CA 91010, USA
e-mail: tlee@smtplink.Coh.org

ANTHONY H. MERRY
The Glycobiology Institute, Department of Biochemistry, South Parks Road,
Oxford OX1 3QU, UK
e-mail: tony@merryone.demon.co.uk

NEIL MURPHY
The Glycobiology Institute, Department of Biochemistry, South Parks Road,
Oxford OX1 3QU, UK
e-mail: neil@bioch.ox.ac.uk

MANFREDO QUADRONI
Protein-Chemistry Laboratory, Universitaetstrasse 16, ETH-Zentrum,
8092 Zurich, Switzerland
e-mail: manfredo.quadroni@bc.biol.ethz.ch

LOUISE ROYLE
The Glycobiology Institute, Department of Biochemistry, South Parks Road,
Oxford OX1 3QU, UK
e-mail: louise@glycob.ox.ac.uk

PAULINE M. RUDD
The Glycobiology Institute, Department of Biochemistry, South Parks Road,
Oxford OX1 3QU, UK
e-mail: pmr@oxglua.glycob.ox.ac.uk

BERNHARD SPENGLER
Institut für Physikalische Chemie, Universität Würzburg, Am Hubland,
97074 Würzburg, Germany
e-mail: bspengler@phys-chemie.uni-wuerzburg.de

DOUGLAS T. STAHL
Division of Immunology, Beckman Research Institute of the City of Hope,
1450 East Duarte Road, Duarte, CA 91010, USA
e-mail: dstahl@smtplink.Coh.org

WERNER STAUDENMANN
Protein Chemistry Laboratory, Universitaetstrasse 16, ETH-Zentrum,
8092 Zurich, Switzerland
e-mail: werner.staudenmann@bc.biol.ethz.ch

DAVID L. TABB
University of Washington, Department of Molecular Biotechnology,
Box 357730, Seattle, WA 98195-7730, USA
e-mail: dtabb@u.washington.edu

JOHN R. YATES III
University of Washington, Department of Molecular Biotechnology,
Box 357730, Seattle, WA 98195-7730, USA
e-mail: jyates@u.washington.edu

1 Mass Spectrometry and the Proteome

Peter James

1.1 The Industrialisation* of Biology

Recently, biological vernacular has been expanded with a series of "omes": the genome (the DNA sequence of an organism), the transcriptome [the messenger RNA (mRNA) expressed at a given time in a cell] and the proteome (the protein equivalent). The latest in the family has been dubbed the metabolome and is a term for all small molecules that are a product of enzymatic and chemical activity within the cell. In contrast to the genome, which is fairly inert, the latter three molecular groups are highly dynamic and vary greatly according to the endogenous and exogenous conditions; they also vary throughout the life cycle of an organism. The human genome, for example, consists of a four-letter nucleotide alphabet organised into a few dozen molecules (the 46 chromosomes). It encodes between 50,000 and 100,000 genes, which can be either directly transcribed (in a 1:1 ratio) or can be recombined into various different combinations by gene rearrangement (for example, T-cell receptors and immunoglobulins). The dynamic expression of these genes as mRNA (the transcriptome), which also uses a four-letter alphabet, can be followed in a quantitative and qualitative manner using binding assays based on DNA arrays (Schena et al. 1995). Alternatively, sequencing methods based on either differential-display polymerase chain reactions (PCRs; Liang and Pardee 1992) or a tagged-DNA approach termed serial analysis of gene expression (Velculescu et al. 1995) can be used. These tools allow one to carry out global experiments, such as systematic analyses of total gene expression following the perturbation of a system (Lashkari et al. 1997; Velculescu et al. 1997).

The key to the development of these large-scale mRNA-expression studies were technological advances in analytical biochemistry, such as the development of PCRs, shotgun sequencing and fluorescently tagged DNA capillary sequencing (Saiki et al. 1985; Adams et al. 1991; Scherer et al. 1999). In order to obtain reproducibility and data accuracy at the high-throughput levels necessary for the assembly of these complex data sets, the process had to be automated by robotics and new algorithms for data assembly, and sequence evaluation had to be developed. The combination made the genome

---

* Synonyms for industrialisation: automation, computerisation, mechanisation, artificial intelligence, cybernetics

sequencing not only possible, but almost routine. A 128-capillary-array DNA sequencer can produce 128,000 bases per run, giving an output of three million bases, equivalent to a bacterial genome per day. The focus of biological problem solving can now move from a reductionist approach to a global approach. Instead of dissecting a process to identify a putative single effector, more subtle analyses based on monitoring the expression of the entire genome can be carried out. The onus of extracting the salient features of a process from this mass of data is dependent on advances in bioinformatics. The huge data sets must be analysed and displayed to allow a scientist to visualise the complex relationships between objects previously thought to be unrelated.

1.2 The Proteome

1.2.1 mRNA and Protein

DNA and mRNA are physico-chemically very homogenous and "easy" to handle and can be amplified by PCR; hence, they are amenable to automation. However, there are several key objections to reducing biological studies to the monitoring of changes in mRNA: (1) the level of mRNA does not allow one to predict the level of protein expression (Anderson and Seilhamer 1997; Gygi et al. 1999b); (2) protein function is controlled by many post-translational modifications; and (3) protein maturation and degradation are dynamic processes that dramatically alter the final amount of active protein, independent of the mRNA level.

In order to be able to correlate mRNA levels with protein expression, modification and activity, there should be a systematic method for separating and visualising the protein components of a cell that allows the : (1) extraction and high-resolution separation of all protein components, including membrane, extreme-pI and low-copy-number proteins; (2) identification and quantification of each component; and (3) comparison, analysis and visualisation of complex changes in expression patterns.

1.2.2 Visualising the Proteome: Two-Dimensional Gel Electrophoresis

Since proteins are vastly more physico-chemically diverse than nucleic acids, a universal separation method is unlikely to be found; this is further compounded by the lack of an amplifying method analogous to PCR. The only partially satisfactory methods for analysing the state of expression of the majority of proteins in a cell are those based on two-dimensional polyacrylamide gel electrophoresis (2D-PAGE, Klose 1975; O'Farrell 1975; Scheele 1975). Proteins are separated in the first dimension according to their isoelectric point, i.e. by migrating to a point in the gel where the pH causes the net charge on the protein to become neutral. In the second dimension, they are separated according to their mobility in a porous gel; this is proportional to the amount of detergent (sodium dodecyl sulphate) bound, which is approximately mass dependent. This methodology allows the resolution of more than 10,000 proteins and can differentiate among many post-translationally modified forms of a protein. The increase in reproducibility that has resulted from the introduction of commercial immobilised pH-gradient first-dimension gels (IPG; Bjellqvist et at. 1982) allows very accurate and quantitative comparative 2D gel mapping. Detailed "proteome maps" can be created with advanced computer imaging programs and can then be analysed by subtractive or cluster methods to find relationships among the protein spots.

The weakness of 2D-PAGE is its inability to deal with certain classes of proteins, mostly highly hydrophobic ones (especially membrane and cytoskeletal proteins) and those with isoelectric points at either extreme of the pH scale (such as acidic, hyper-phosphorylated proteins and alkaline, DNA-binding proteins). There are also problems with quantification due to the low dynamic range of stains. Silver staining, the most sensitive commonly used stain, is linear only between 0.04 ng/mm^2 and 2 ng/mm^2. The use of narrow pH-range IPG gels and sample loading via re-hydration (Schupbach et al. 1991; Rabilloud et al. 1994) can partially compensate for this problem by allowing much larger amounts of protein to be loaded, thus extending the sensitivity range. A detailed account of the art and science of 2D-PAGE can be found in an accompanying book in this series.

1.3 Mass Spectrometry and Proteomics

1.3.1 General Requirements

The basic requirements of proteome analysis are: a wide dynamic-detection range, high-throughput and high-confidence protein identification, protein quantification, the ability to deal with multiple proteins in a single spot and the ability to identify post-translational modifications. Mass spectrometry (MS) can fulfil all of these requirements.

1.3.2 Sensitivity

Cells exhibit a large dynamic range of expression – between one and 10^8 copies of a protein per cell. The sensitivity range required can be roughly calculated in this way: a culture plate contains 10^6 cells; thus, one needs to cover the range 60 amol to 60 nmol. Since the sensitivity range of silver staining extends only to the low-femtomole range, either new fluorescent dyes or radioactivity must be used to visualise the low-attomole range. Since a single cell may express

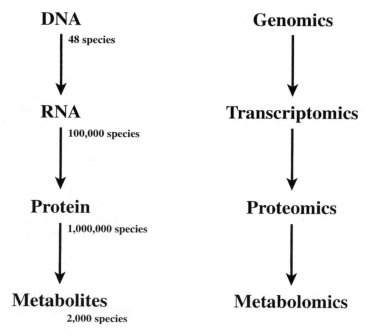

Fig. 1.1. The different types of molecules and the approximate number of different types (in humans), and their relationships to the new biological "omics" terms

10,000 or more proteins at the same time, and since (on average) a single gene produces ten spots, one can visualise only a small potion of the data, even using the most sensitive staining and the highest-resolution gels. The sensitivity and mass range of MS instruments has increased in several quantum leaps during the past two decades (Fig. 1.1; Chap. 3). This has been due to the introduction of new ionisation methods like fast atom bombardment (Barber et al. 1981), matrix-assisted laser desorption and ionisation (MALDI; Karas and Hillenkamp 1988) and electrospray ionisation (ESI; Fenn et al. 1989). The designs of the most commonly used MS instruments for analysing proteins and peptides have not changed greatly, and the principles of their operation are described in Chap. 2 (triple-quadrupole and ion-trap MS) and Chap. 3 (time-of-flight MS). Commercially available instruments can provide high-quality sequence spectra in the low-femtomole region (Morris et al. 1996), though research instruments can attain low-attomole to high-zeptomole sensitivities (Andren et al. 1994; Valaskovic et al. 1996).

1.3.3 Sample Handling

New methods must be developed for protein detection on 2D gels if one is to match these levels. Another problem is how to digest the proteins for subsequent analysis. Most endoproteases have a K_m in the micromolar range (i.e. they exhibit 50% maximum velocity with proteins in the low picomole per microlitre range), whilst the concentration of protein in a silver-stained spot is approximately one thousandth of this. In order to be able to digest low-abundance proteins, they must first be concentrated, and miniaturised methods for sample concentration have recently been described for DNA using a device fabricated on a silicon chip (Khandurina et al. 1999). Once digested, the peptides can either be directly analysed as a mixture or can be separated via capillary high-performance liquid chromatography (Chap. 4) or solid-phase extraction capillary-zone electrophoresis (Chap. 5) coupled directly to a mass spectrometer. The amount of data that can be extracted by these on-line methods is much higher and, since the peptides are concentrated into smaller volumes, a greater level of sensitivity can be achieved.

1.3.4 High-Throughput Protein Identification

Since up to 10,000 protein spots can be visualised on a single gel, any technique used for the analysis and identification must, by definition, be high throughput. The development of the ESI and MALDI inlet techniques made possible the development of MS-based protein-identification methods. Recently, a very rapid method of sample delivery from 96-well microtitre plates for automated ESI-MS analysis (Felten et al. 1999) has been described. The trend is toward miniaturised sample-handling devices that automate all the steps from an excised protein spot to delivery into the mass spectrometer (Jedrzejewski et al. 1999). Chapters 4 and 5 describe methods for the subsequent automation of data accumulation. The use of peptide masses and peptide-fragmentation spectra to identify proteins in an automated, high-throughput fashion are the subjects of Chapters 6 and 7, respectively. If a protein is not present in a database, individual peptides may be characterised by fragmentation in a mass spectrometer. The resulting MS/MS spectra can be interpreted either manually (as described in Chap. 8) or with the help of computer programs (which is dealt with in Chap. 9).

1.3.5 Quantification

Until recently, protein quantification usually involved amino acid analysis or, less precisely, the intensity of staining on a 2D gel. Recently, isotopic-labelling methods that allow the quantitative determination of relative changes in protein amounts or modification levels by MS have been described (Oda et al. 1999). The control cells are grown in a normal medium, whilst the experimental cells are grown in a medium with a stable isotope, such as deuterium or ^{15}N. The cell extracts are mixed and separated by 2D gel electrophoresis. The protein of interest is excised and digested, and the change in expression or modification level is determined by comparing the isotope ratios of the labelled and non-labelled peptides.

A very promising alternative to 2D gel analysis as a comprehensive method for comparative proteomics was recently described by the group of Aebersold and is called isotope-coded affinity tagging (Gygi et al. 1999a). The conceptual basis of the technique is illustrated in Fig. 1.2. Essentially, whole-cell protein extracts are digested and labelled with either a "light" or "heavy" deuterated-biotin label that has a thiol-specific reactive group. The mixture is digested, and the biotinylated peptides are recovered using an avidin affinity column. This much simpler peptide mixture is then analysed via LC-MS using a reverse-phase column. The isotopically labelled pairs of peptides elute almost simul-taneously, and the isotope peak ratios can give the relative amounts of each. It is then necessary to perform MS/MS analysis of the peptides to determine their sequence and thus identify the parent protein. This technique is broadly applicable and represents the most viable alternative to 2D-PAGE available today.

1.3.6 Post-Translational Modifications

Of the over 300 naturally occurring post-translational modifications, most are accessible to analysis via MS. The two most commonly occurring and important modifications, phosphorylation and glycosylation, are dealt with in Chapters 10 and 11, respectively.

1.4 Aims of the Book

Despite the development of highly automated protein handling and identifica-tion methodologies, there will always be a need to understand the basic prin-ciples underlying these techniques. This book aims to provide a foundation to aid understanding of the hardware and software tools that are currently being developed. MS has already become an indispensable tool for the biologist, and its utility and breadth of application are rapidly increasing. The chapters provide both an up-to-date review of the current literature and discuss prac-tical aspects of transferring the methods to the laboratory. I hope that, by pre-senting the principles in the context of the biological problems they are intended to solve, this will make learning more pleasurable and will provide an up-to-date view of laboratory procedures.

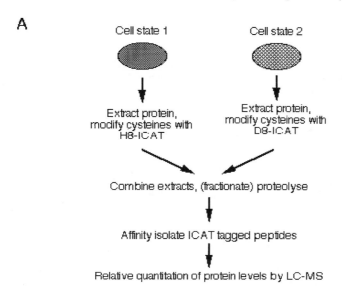

A

Cell state 1   Cell state 2

Extract protein,
modify cysteines with
H8-ICAT

Extract protein,
modify cysteines with
D8-ICAT

Combine extracts, (fractionate) proteolyse

Affinity isolate ICAT tagged peptides

Relative quantitation of protein levels by LC-MS

B

Isotope-Coded Affinity Tag

HN   NH

H/D   H/D
H/D
O   O
N   O   N   I
H   H/D   H/D   H
H/D   H/D

S

Biotin   Linker -Heavy (all D) or light (all H)   Thiol specific group

Fig. 1.2A,B. Isotope-coded affinity tagging as an alternative to two-dimensional polyacrylamide-gel electrophoresis (2D-PAGE). The protein extracts from cells grown under different conditions are specifically labelled with the biotin isotope label at cysteine residues. The extracts are digested and mixed; labelled peptides are then selectively isolated by affinity chromatography. The recovered peptides are analysed by high-performance liquid chromatography/mass spectrometry, and the change in expression is determined via the relative isotope levels. The reduction in complexity introduced by the affinity purification step allows one to bypass the need for a 2D-PAGE separation

References

Adams MD, Kelley JM, Gocayne JD, Dubnick M, Polymeropoulos MH, Xiao H, Merril CR, Wu A, Olde B, Moreno RF, Venter JC (1991) Complementary DNA sequencing: expressed sequence tags and human genome project. Science 252:1651–1656

Anderson L, Seilhamer J (1997) A comparison of selected mRNA and protein abundances in human liver. Electrophoresis 18:533–537

Andren PE, Emmet MR, Caprioli RM (1994) Micro-electrospray: Zeptomole/Attomole per micro-liter sensitivity for peptides. J Am Mass Spectrom 5:867–869

Barber M, Bordoli RS, Sedgwick RD, Tylor AN (1981) Fast atom bombardment of solids as an ion source in mass spectrometry. Nature 293:270–275

Bjellqvist B, Ek K, Righetti PG, Gianazza E, Gorg A, Westermeier R, Postel W (1982) Isoelectric focusing in immobilized pH gradients: principle, methodology and some applications. J Biochem Biophys Methods 6:317–339

Felten C, Foret F, Karger B (1999) A simple procedure for automated high-throughput infusion ESI-MS through direct coupling of 96-well microtiter plate to sub-atmospheric pressure ESI interface. The 47th ASMS Conference on Mass Spectrometry and Allied Topics, Dallas, Texas, 1999. CD-ROM. The American Society of Mass Spectrometry

Fenn JB, Mann M, Meng CK, Wong SF, Whitehouse CM (1989) Electrospray ionization for mass spectrometry of large biomolecules. Science 246:64–71

Gygi SP, Rist B, Gerber SA, Turecek F, Gleb MH, Aebersold R (1999a) Quantitative analysis of complex protein mixtures using isotope-coded affinity tags. Nat Biotechnol 17:994–999

Gygi SP, Rochon Y, Franza BR, Aebersold R (1999b) Correlation between protein and mRNA abundance in yeast. Mol Cell Biol 19:1720–1730

Jedrzejewski P, Zhang B, Karger B (1999) High-throughput proteomic sample preparation on microfabricated devices for sensitive MS detection and identification. The 47th ASMS Conference on Mass Spectrometry and Allied Topics, Dallas, Texas, 1999. CD-ROM. The American Society of Mass Spectrometry

Karas M, Hillenkamp F (1988) Laser desorption ionization of proteins with molecular masses exceeding 10,000 daltons. Anal Chem 60:2299–2301

Khandurina J, Jacobson SC, Waters LC, Foote RS, Ramsey JM (1999) Microfabricated porous membrane structure for sample concentration and electrophoretic analysis. Anal Chem 71:1815–1819

Klose J (1975) Protein mapping by combined isoelectric focusing and electrophoresis of mouse tissues. A novel approach to testing for induced point mutations in mammals. Humangenetik 26:231–243

Lashkari DA, DeRisi JL, McCusker JH, Namath AF, Gentile C, Hwang SY, Brown PO, Davis RW (1997) Yeast microarrays for genome wide parallel genetic and gene expression analysis. Proc Natl Acad Sci USA 94:13057–13062

Liang P, Pardee AB (1992) Differential display of eukaryotic messenger RNA by means of the polymerase chain reaction. Science 257:967–971

Morris HR, Paxton T, Dell A, Langhorne J, Bordoli RS, Hoyes J, Bateman RH (1996) High sensitivity collisionally-activated decomposition tandem mass spectrometry on a a novel quadrupole/orthogonal-acceleration Time-of-Flight mass spectrometer. Rapid Commun Mass Spectrom 10:889–896

Oda Y, Huang K, Cross FR, Cowburn D, Chait BT (1999) Accurate quantitation of protein expression and site-specific phosphorylation. Proc Natl Acad Sci USA 96:6591–6596

O'Farrell PH (1975) High resolution two-dimensional electrophoresis of proteins. J Biol Chem 250:4007–4021

Rabilloud T, Valette C, Lawrence JJ (1994) Sample application by in-gel rehydration improves the resolution of two-dimensional electrophoresis with immobilized pH gradients in the first dimension. Electrophoresis 15:1552–1558

Saiki RK, Scharf S, Faloona F, Mullis KB, Horn GT, Erlich HA, Arnheim N (1985) Enzymatic amplification of beta-globin genomic sequences and restriction site analysis for diagnosis of sickle cell anemia. Science 230:1350–1354

Scheele GA (1975) Two-dimensional gel analysis of soluble proteins. Charaterization of guinea pig exocrine pancreatic proteins. J Biol Chem 250:5375–5385

Schena M, Shalon D, Davis RW, Brown PO (1995) Quantitative monitoring of gene expression patterns with a complementary DNA microarray. Science 270:467–470

Scherer JR, Kheterpal I, Radhakrishnan A, Ja WW, Mathies RA (1999) Ultra-high throughput rotary capillary array electrophoresis scanner for fluorescent DNA sequencing and analysis. Electrophoresis 20:1508–1517

Schupbach J, Ammann RW, Freiburghaus AU (1991) A universal method for two-dimensional polyacrylamide gel electrophoresis of membrane proteins using isoelectric focusing on immobilized pH gradients in the first dimension. Anal Biochem 196:337–343

Valaskovic GA, Kelleher NL, McLafferty FW (1996) Attomole protein characterization by capillary electrophoresis-mass spectrometry. Science 273:1199–1202

Velculescu VE, Zhang L, Vogelstein B, Kinzler KW (1995) Serial analysis of gene expression. Science 270:484–487

Velculescu VE, Zhang L, Zhou W, Vogelstein J, Basrai MA, Bassett DE, Hieter P, Vogelstein B, Kinzler K (1997) Characterization of the yeast transcriptome. Cell 88:243–251

2 Basics of Triple-Stage Quadrupole/Ion-Trap Mass Spectrometry: Precursor, Product and Neutral-Loss Scanning. Electrospray Ionisation and Nanospray Ionisation

DAVID ARNOTT

2.1 Introduction

Mass analysis based on the motion of ions in an oscillating electric field has evolved over a period of more than 30 years (Finnigan 1994). Quadrupole ion traps and triple-stage quadrupole (TSQ) instruments are among the most versatile mass spectrometers yet developed, primarily because of their power in the structural analysis of molecules through tandem mass spectrometry (MS/MS). Both types of instruments are capable of product-ion scanning – the isolation of a chosen precursor ion and the detection of its product ions following collision-induced dissociation (CID). Ion traps can perform multiple stages of precursor isolation and fragmentation in a single experiment (MS^n). Triple-quadrupole instruments can also be used in other scanning modes. Precursor-ion scans detect ions that fragment to produce a selected product ion, and neutral-loss scans identify ions that, upon fragmentation, eject a neutral species with a specified mass.

Although proteomics is a young field and its full scope is uncertain, several areas of research have already appeared. Much of its growth has stemmed from its ability to identify proteins by comparing their proteolytically derived fragments with those expected from each entry in the rapidly expanding primary-sequence databases (Henzel et al. 1993; James et al. 1993; Mann et al. 1993; Yates et al. 1993). In addition to peptide mass, sequence information obtained from MS/MS can be used for this purpose; these data can be obtained from both triple-quadrupole and ion-trap instruments (Mann and Wilm 1994; Yates et al. 1995; Arnott et al. 1998a). An extension of this technique is the de novo sequencing of currently unknown proteins (Wilm et al. 1996a). The need for this experiment may diminish as the genomic sequencing of model organisms is completed, but it will not disappear. Furthermore, proteomics is likely to be valuable in the study of post-translational modifications. The neutral-loss and precursor-ion scanning modes of the triple-quadrupole mass spectrometer can be used to specifically detect modified peptides, and the MS^n capability of ion traps facilitates the structural characterisation of the modifications.

A rich body of literature exists for both quadrupole and ion-trap MS. This chapter provides a brief overview of the basic principles and modes of operation that make these instruments useful for peptide analysis. More comprehensive reviews of quadrupoles can be found by Dawson (1986), and reviews of ion traps can be found in the *Practical Aspects of Ion-Trap Mass Spectrometry* series edited by March and Todd (1995).

2.2 Fundamental Principles

2.2.1 Mass Analysis

The operation of all quadrupole and quadruple ion-trap instruments is based on the motion of ions in oscillating (radio-frequency, RF) electric fields; this was elaborated by Wolfgang Paul in 1960, for which he was awarded the Nobel Prize for physics in 1989. The basis for mass analysis is that, in an RF field, an ion experiences a force proportional to its distance from the centre of the field. This is easily visualised in one dimension as an ion between oppositely charged electrodes is accelerated away from one towards the other. If the polarity of the electrodes is switched, the ion will be accelerated in the opposite direction and, if the amplitude and frequency at which the field oscillates are sufficient, the ion will be confined between the electrodes. In directions perpendicular to the electrodes, of course, the motion of the ion is not constrained. A two-dimensional quadrupolar field, in which focusing is applied by two pairs of electrodes (Fig. 2.1A), confines ions along two axes while allowing them to drift along a third. Ions can be trapped in three dimensions, in the volume defined by a ring electrode and two end-cap electrodes of an ion trap (Fig. 2.1B).

The precise behaviour of ions, and the means by which they can be manipulated in a mass-dependant manner, can be derived from the time-dependant potential θ at any point in the electric field:

$$\theta = [U - V\cos(\omega t)](x^2 + y^2 - z^2)/2r_0^2 \tag{1}$$

where r_0 is the radius of the ring electrode (or the distance between rods in a quadrupole), V is the amplitude of the RF signal with frequency ω, and U is the direct current (DC) potential applied between electrodes. The motion of an ion with mass m and charge z is then determined by three differential equations (one for each direction: x, y and z) of the form known as the Mathieu equation. For example, along the x-axis:

$$\frac{\partial^2 x}{\partial t^2} + \left(\frac{2z}{mr_0^2}\right)(U - V\cos(\omega t))x = 0 \tag{2}$$

The most common way to express solutions to these equations is to collect terms into two parameters a and q*, where a is proportional to the DC poten-

---

* The parameters a and q are subscripted to indicate the direction in which they apply, i.e. a_z and q_z for the axial direction in ion traps, or a_x and q_x (which equal $-a_y$ and $-q_y$, respectively) in quadrupoles.

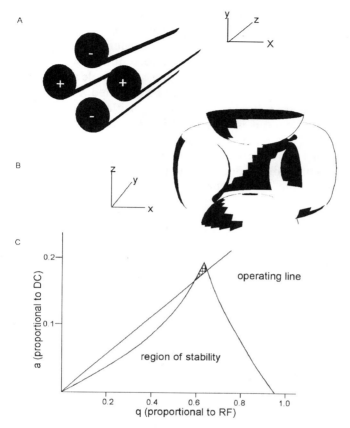

Fig. 2.1A,B. Principles of mass analysis. **A** Two-dimensional quadrupole mass filter. Radio-frequency and direct-current voltages applied between pairs of rod-shaped electrodes (ideally of hyperbolic cross section) focus ions in the x and y dimensions, allowing them to drift along the z-axis. **B** Trapping ions in three dimensions. A ring electrode and two end-cap electrodes constrain the motion of ions radially (in the x–y plane) and axially (along the z-axis). **C** The region of the stability diagram used by commercial quadrupole and ion-trap instruments. The operating line for scanning a mass spectrum in quadrupoles is indicated; only ions with q values in the shaded area are passed

tial, q is proportional to RF potential and all other terms except m/z are constant.

$$a_z = \frac{-8U}{(m/ze)r_0^2\omega^2}, q_z = \frac{4V}{(m/ze)r_0^2\omega^2} \qquad (3)$$

Therefore, at any values of U and V, every ion will have a unique value of a and q and can be described as having a stable or unstable trajectory. Combinations of a and q that have stable solutions have been plotted in stability diagrams; Fig. 2.1C shows the region of stability employed in commercial quadrupole and ion-trap instruments.

Once the stability diagram is understood, the operation of quadrupole mass spectrometers is straightforward. The RF and DC potentials on the quadrupole rods are set so that only ions with a small range of m/z values fall within the apex of the stability diagram. As the voltages are ramped upwards, ions with successive m/z values will pass down the length of the quadrupole with stable trajectories and be detected, whereas all others will be deflected. The ratio of the RFs and DC voltages that are scanned during the acquisition of a spectrum is known as the operating line and is illustrated in Fig. 2.1C. This directly results in two characteristics of quadrupole instruments. First, at any given time during a scan, the vast majority of ions are discarded; thus, quadrupoles are properly described as mass filters. Second, the mass resolution obtained is a function of the operating line. If the slope is lowered, more ions covering a larger range of m/z pass through, increasing the signal detected but reducing the resolution (Watson 1997).

The operation of ion-trap instruments is somewhat more complex, because several means of mass analysis can be employed; nevertheless, the stability diagram is instructive. These instruments generally operate with no DC potential applied, so ions fall along the q_z axis from high m/z to low m/z. Those with q values below 0.91 (the right edge of the stability region) remain in the trap, while those with q values above this limit collide with the end-cap electrodes or are ejected if holes have been drilled in the end caps. If the RF voltage is ramped upwards, ions will have successively higher q values, and each will ultimately reach the edge of the stability region and be ejected, at which point it will impinge on an external detector. This mode of scanning, known as mass-selective instability, was the first practical means of using ion traps as mass spectrometers, and its invention made commercial ion-trap instrumentation possible (Stafford et al. 1984). A related means of mass analysis takes advantage of the fact that, at any given q value, an ion will oscillate within the trap at a defined frequency. If an additional RF voltage is applied between the end-cap electrodes at that same frequency, the ion will increase its amplitude of motion and, if the supplementary voltage is high enough, the ion will be ejected from the trap. Therefore, if the main RF voltage is increased to bring successive ions into resonance with the excitation voltage, ions will be ejected and detected sequentially according to their masses. The frequency is generally chosen to eject ions near the edge of the stability region (i.e. $q_z = 0.904$). It is also possible to eject ions at lower q values, effectively increasing the mass range that can be scanned (Louris et al. 1990). In practice, resonance excitation yields resolution and signal-to-noise ratios superior to those obtained via mass-selective instability; it is the basis for mass analysis in commercial instruments, such as the Finnigan MAT LCQ and the Bruker ESQUIRE (March 1998).

2.2.2 Collision-Induced Dissociation

A key to the versatility of triple-quadrupole and ion-trap instruments is the ability to perform MS/MS. Valuable structural information can be gathered by subjecting a precursor ion (isolated in an initial stage of MS) to CID, followed by a second stage of mass analysis on the product ions. In CID (also correctly called collisionally activated dissociation), ions gain internal energy through collisions with an inert gas; they subsequently fragment via a process of uni-molecular decomposition. Fragmentation requires that sufficient energy be deposited in the ion so that the vibrational energy in a given bond exceeds its bond strength. Triple quadrupoles and ion traps employ low-energy collisions, typically less than 30 eV (lab frame) for triple quadrupoles, and even less for ion traps, as opposed to the hundreds to thousands of electron volts deposited in magnetic-sector or tandem time-of-flight instruments (McCormack et al. 1993). For peptides undergoing low-energy CID, internal energy is usually deposited in a series of collisions before enough energy (which is distributed among all vibrational modes) accumulates. Low-energy CID is particularly useful in the analysis of peptides, because fragmentation occurs frequently at amide bonds, leading to product ions characteristic of the peptide's sequence (Hunt et al. 1986; Tang et al. 1993).

Subsequent chapters of this book discuss both the interpretation of CID spectra and automated techniques for matching spectra to peptide sequences in protein databases, so only a few features of these spectra need to be men-tioned here. Fragmentation of peptides at amide bonds is so random that extensive series of ions, each differing in mass by one amino acid, are usually obtained for singly, doubly and (often) triply charged precursors with molec-ular masses less than 3000 Da. Larger or more highly charged precursors are not as cooperative, either because not enough energy can be deposited in collisions to induce fragmentation (Griffin and MacAdoo 1993) or because the presence of charged amino acids in the middle of the peptide leads to fragmentation at only a few preferred bonds, so the spectra are not as rich and informative. Furthermore, CID can produce fragments that contain the N-terminal portion of the peptide (designated as a, b, c or d ions, depending on the bond cleaved) or fragments that contain the C-terminus (designated w, x, y or z ions). Roepstorff and Fohlman (1984) first proposed the nomenclature, and Biemann's modification of this system is used here (Biemann 1988). In low-energy conditions a, b and y ions are the predominant fragments observed. This is in contrast to high-energy CID patterns, in which more energetic collisions allow access to more fragmentation mechanism pathways. Product ions can fragment further to produce "internal ion series", which do not contain either terminus. A special class of internal ion is the immonium ion, found in the low-mass region of CID spectra. These ions are characteris-tic of the peptide's amino acid composition. Ions can also undergo neutral losses of small molecules, such as ammonia or water; such losses are charac-teristic of the structures of particular amino acids. The way this experiment is performed and the scan modes that can be accessed are quite different in ion traps than in triple quadrupoles and are described in detail below for both instruments.

2.2.3 Electrospray Ionisation

Electrospray ionisation (ESI) is the most commonly used source of ions for the study of peptides and proteins in both quadrupole and ion-trap mass spectrometers. First described as a viable ionisation method for MS by Fenn and co-workers (Whitehouse et al. 1985), ESI produces ions through the application of voltage to a stream of liquid emitted from the opening in a tube or capillary. A cone of highly charged droplets is sprayed from the opening at atmospheric pressure and is sampled by the mass spectrometer. The droplets shrink and desorb the charged analytes as they pass through several stages of differential pumping. ESI has the virtues of sampling directly from liquid phase, imparting relatively little internal energy to the analytes (so that intact macromolecules can be detected) and efficiently bringing polar and charged molecules to the gas phase. Polypeptides containing more than five or six residues are almost always observed as multiply protonated species with charge states increasing in proportion to the size and the number of basic amino acids present. Proteins with molecular masses far in excess of the mass range possessed by a mass spectrometer can usually be detected, because they bear enough charges to produce an "envelope" of peaks with m/z less than 2000 (Smith et al. 1990).

One reason ESI has found such extensive use is its compatibility with on-line separations by chromatography or capillary electrophoresis. It is often advantageous to resolve the components of a mixture and, because ESI is sensitive to the concentration rather than the total quantity of analyte, chromatographic concentration of samples improves detection limits relative to direct infusion into the source (Davis et al. 1995). Liquid chromatography (LC) and capillary electrophoresis are applied to diverse experiments, including a wide range of flow rates, from nanolitres per minute to greater than 1 ml/min; ESI sources have evolved to accommodate such different conditions. At the higher flow rates (microlitre per minute and higher), a nebulising gas usually assists the formation of droplets (Covey et al. 1988) whereas, at lower flow rates, the liquid is often emitted from a narrow bore or tapered needle with an inner diameter of less than 50 μm (Wahl et al. 1994; Davis et al. 1995). ESI with flow rates of hundreds of nanolitres per minute is generally described as micro-electrospray.

The logical extension of spraying from ever-smaller ESI emitters with the lowest achievable stable flow rates is known as nanospray (Wilm and Mann 1994). This technique is usually performed using glass capillaries with tips pulled to inner diameters of less than 5 μm, and flow rates of a few nanolitres per minute are achievable; samples with volumes of 1–2 μl can be analysed over the course of an hour or longer. Such extended analysis times permit signal averaging to enhance the spectral quality of low signal-to-noise samples, and many MS/MS experiments can be performed during a single experiment. Nanospray represents nearly ideal electrospray conditions, including maximum ionisation efficiency and the transport of sample into the mass spectrometer (Mann and Wilm 1996). However, interfacing nanospray with chromatography is problematic, so sample preparation must usually be done off-line; thus, the concentration-sensitivity advantage of LC-MS is foregone.

2.3 Triple-Quadrupole MS

2.3.1 Instrument Overview

Use of the quadrupole as a mass analyser was described in Section 2.1. The triple quadrupole, as the name suggests, is constructed as three quadrupoles in series, separated by focusing lenses; they are preceded by an ion source and followed by a detector, as illustrated in Fig. 2.2A. The use of an ESI source introduces the challenge of transporting ions formed at atmospheric pressure into the 10^{-6}-Torr analyser region of the instrument. This is accomplished via several stages of differential pumping. Depending on the design of the source, ions can be brought in with or without desolvation/curtain gases, directly through an orifice in the vacuum manifold or through a heated capillary. Various arrangements of lenses and skimmer cones can be used to desolvate and focus ions into the first quadrupole (Q1) while pumping away solvent and atmospheric gases. Whatever the case, there is necessarily a region of relatively high pressure through which ions travel, during which they can collide with gas molecules and possibly undergo fragmentation, known as "in-source CID". This can be a problem when analysing highly labile molecules but can be used to carry out more complex structural studies than would otherwise be possible with a TSQ instrument.

Mass analysis can be performed using either Q1 or Q3; Q2 is operated in RF-only mode, meaning it can transmit ions having a wide range of m/z values. This quadrupole is used as part of a collision cell; during MS/MS experiments, an inert gas with which precursor ions passed by Q1 can collide is bled into the cell. Argon is normally used as the collision gas, although xenon is occasionally employed to impart higher energy per collision (useful for higher-mass peptides, which otherwise resist fragmentation). Several instrumental parameters influence the efficiency of CID. The pressure in the collision cell dictates the number of collisions an ion will undergo as it passes through Q2. The "offset" of the voltage on Q2, which is the potential difference accelerating an ion through the collision cell, determines the energy imparted by each collision. In practice, it is usually easiest to use a fixed collision-gas pressure and

A. Triple Quadrupole Schematic

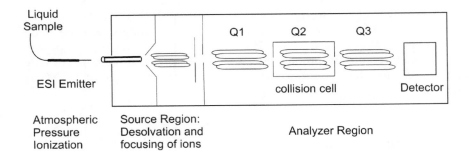

B. Ion Trap Schematic

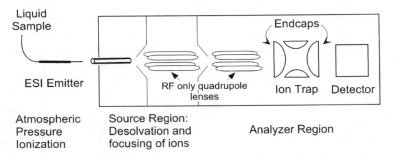

Fig. 2.2A,B. Components of mass spectrometers. Diagrams of **A** a triple-quadrupole instrument and **B** an ion-trap mass spectrometer. The schematics are generic; "real" instruments vary in the specifics of the ion source, ion optics, etc

choose collision offset voltages for optimal fragmentation of a given peptide. The best conditions depend on the mass of the peptide, its charge state and, to some extent, its sequence.

2.3.2 Product-Ion Scans

Product-ion scans are the most commonly employed type of MS/MS for peptide-sequence analysis. Conceptually, they are also the easiest to understand. A precursor ion, chosen on the basis of prior knowledge of the sample (from a previous Q1 or Q3 scan) is isolated by setting Q1 to pass only that ion. CID is carried out in Q2, and all product ions are mass analysed in Q3. The rich structural information this type of scan can provide is illustrated in

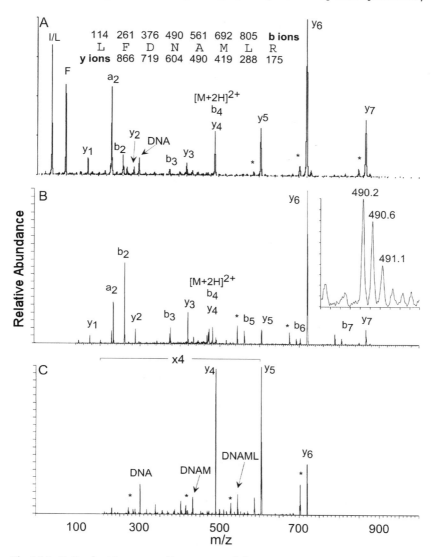

Fig. 2.3A–C. Product-ion scans of human growth-hormone peptide T2. The mass/charge ratio of $[M + 2H]^{2+}$ is 980.2. The peptide sequence and predicted b and y ions are shown in the *inset*. Ions are labelled accordingly; *asterisks* indicate neutral loss of H_2O or NH_3. **A** Triple quadrupole. **B** Ion trap. **C** MS^3 on an ion trap. After dissociation of the m/z = 980.2 precursor, the y_6 ion at m/z = 719 was isolated and fragmented

Fig. 2.3A. The doubly charged ion of peptide LFDNAMLR yields a product-ion scan containing a complete series of y ions, allowing the entire sequence of the peptide to be determined, except that leucine and isoleucine cannot be distinguished from each other, since they have identical masses. A possible ambiguity due to the fact that ions b_4 and y_4 have the same nominal mass as the $[M + 2H]^{2+}$ precursor is resolved by the presence of the internal ion with m/z = 301, representing the sequence Asp–Asn–Ala (resulting from further decomposition of the y_6 ion). This spectrum is characteristic of tryptic peptides because, having a basic amino acid at the C-terminus, the formation of y ions is favoured (McCormack et al. 1993). Of the N-terminal fragments, b_2, a_2 and b_3 are detected, supplying further confirmatory evidence of the sequence. Other types of fragment ions described in Section 2.2 are also present. Two prominent immonium ions with m/z = 86 and m/z = 120 are characteristic of leucine/isoleucine and phenylalanine, respectively. Lastly, several y ions undergo neutral losses of ammonia and are indicated with *asterisks* in Fig. 2.3A.

As explained in Section 2.1, quadrupoles exhibit a compromise between sensitivity and resolution. Because fragmentation of precursor ions distributes charge among many product ions, and because of losses due to the collisional scattering of ions, more sample is usually required to produce a product-ion scan than a simple Q1 or Q3 scan. Therefore, the resolution of Q3 scans is often lowered in order to boost the signal-to-noise ratio. For the same reason, the Q1 "window" used to pass precursor ions is usually set several mass units wide. This improves limits of detection and allows the full isotopic envelope of a precursor ion to undergo CID, so product ions have "normal" isotopic distributions. There may be, however, situations in which it is advantageous to select a precursor with unit resolution; product ions then contain only the ^{12}C isotopes, allowing two different fragment ions differing by only one or two mass units to be clearly distinguished.

The choice of collision energy is important in any CID experiment, and its effect is most readily apparent in product-ion scans. Without sufficient energy, a precursor ion does not fragment appreciably, and an uninformative spectrum is the result. An excessive collision offset is little better, however. Since ions in the collision cell undergo multiple collisions with the CID gas, product ions can undergo further decomposition to lower mass fragments. If the energy imparted is high enough, the spectrum will contain only low-mass immonium and internal fragment ions, without extensive b or y ion series. If the sample is introduced by direct infusion or nanospray, collision energy can be easily adjusted for each precursor ion. If CID is performed during the elution times of LC-MS or capillary electrophoresis-MS experiments, however, the collision energy must be set in advance. Although attempts have been made to calculate the optimal energy for peptides with any given m/z ratio (Haller et al. 1996), most operators use settings gained from experience. An alternative is to automate data acquisition, taking several scans of each precursor at arbitrary "low", "medium" and "high" collision energies.

2.3.3 Precursor-Ion Scans

In the product-ion scan, discussed above, Q1 is set to transmit ions of a designated mass, while Q3 is scanned. For precursor-ion scanning, the reverse is true. As Q1 is scanned, successive precursor ions enter the collision cell and decompose. Because Q3 is set to transmit ions of a chosen m/z, the detector only registers a signal if a precursor ion transmitted by Q1 fragments to form the selected ion. In this way, ions sharing a structural feature can be detected and resolved from other components of the sample. CID spectra of peptides that contain tyrosine, for example, usually exhibit a prominent immonium ion of m/z = 136. A precursor-ion scan on a mixture of peptides with Q3 set to transmit ions at mass = 136 selectively detects tyrosine-containing peptides. Figure 2.4 demonstrates this experiment. A tryptic digest of human growth hormone (hGH) should, in principle, produce 21 different peptides although, in practice, some are undetected (fragment 17, for example is the amino acid lysine), and other peptides representing incomplete digestion or modifications (such as oxidation or de-amidation) are sometimes encountered. Figure 2.4A shows the Q1 spectrum of an entire hGH digest. Most of the expected peptides are present but, due to multiple charging and the number of peptides in the mixture, the spectrum is complex, with numerous closely spaced peaks. A precursor-ion scan for m/z = 136 (Fig. 2.4B) is considerably simpler, containing only five significant peaks. Each of these represents a tyrosine-containing peptide. Two peaks are related to tryptic peptide 14, and peptide T4 appears as a minor doubly charged ion in addition to the dominant triply charged form. Only one expected tyrosine-containing peptide, T10, is not observed.

The success of a precursor-ion scan depends on three factors. First, for obvious reasons, the precursor ion must be present in sufficient abundance for a signal to be detected following the losses inherent in a CID experiment. Second, the precursor ion must fragment efficiently to form the diagnostic product ion. The m/z = 136 ion, for example, is observed as a large peak in the spectra of most tyrosine-containing peptides. In the example given, a high collision energy was chosen to drive fragmentation to the fullest extent, maximising the abundance of immonium ions. A less stable product ion, however, would require a more optimised collision energy in an attempt to form the fragment without causing it to decompose further. Finally, the chosen product ion must be fairly specific for the class of molecules one wishes to detect. A diagnostic ion with the same mass as a common contaminant or a component of the electrospray chemical background would, therefore, be a poor choice.

Although less often performed than product-ion scanning, the precursor-ion experiment has several powerful applications. A key role of proteomics is likely to be the elucidation of post-translational modifications, and such modifications frequently have characteristic product ions. N-linked glycopeptides, for example, fragment into ions with m/z = 204 (N-acetyl-hexosamine; HexNAc); complex structures fragment to yield ions with

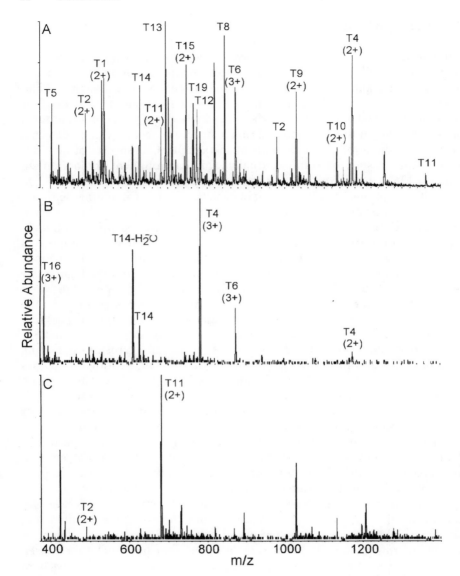

Fig. 2.4A–C. Scan modes of triple quadrupoles illustrated with a tryptic digest of human growth hormone. **A** Q1 scan. Peaks corresponding to expected tryptic fragments are labelled. **B** Precursor ion scan for tyrosine-containing peptides. Ions that fragment to form the m/z = 136 immonium ion are detected. **C** Neutral-loss scan for peptides containing methionine. Ions that fragment to lose 24 mass units (loss of CH_3SH from doubly charged precursors) are detected

m/z = 366 (HexHexNAc). Sialylated glycopeptides can be recognised by an ion at m/z = 292 (Huddleston et al. 1993a). Phosphopeptides analysed in negative-ion mode fragment into products with m/z = 79 and m/z = 63 (PO_3^- and PO_2^-, respectively; Huddleston et al. 1993b). Product-ion scans have also been used in nanospray experiments in which low-abundance peptides obscured by the chemical background could be detected by precursor-ion scans for peptide immonium ions (Wilm et al. 1996b).

2.3.4 Neutral-Loss Scans

Least frequently used for protein and peptide analysis is the neutral-loss scan. When an ion decomposes into two fragments, one or both (if the precursor was multiply charged) can bear a charge and will be detected as a peak in the mass spectrum. If one of the fragments is not charged (always the case for singly charged precursors), then it will not appear as a peak in the spectrum, but the product ion that is detected will differ from the precursor ion by the mass of the neutral species. Such "neutral losses" are common features of CID spectra. Ammonia is often lost from b ions and ions containing arginine; water is lost from numerous species, including y ions and ions containing glutamic acid, aspartic acid, serine and threonine (Ballard and Gaskell 1993). In low-energy collisions, a ions are predominantly produced via the neutral loss of carbon monoxide from b ions.

Neutral losses characteristic of a structural feature can be detected by scanning both Q1 and Q3, with the Q3 scan lagging behind Q1 by a fixed mass. For example, in order to detect only precursor ions that decompose by losing ammonia, Q3 would have an offset of 17 mass units. At any point during the scan, a precursor ion transmitted through Q1 decomposes in Q2; it will have the correct mass to pass through Q3 and be detected only if it undergoes a neutral loss of 17 mass units. This is demonstrated in Fig. 2.4C using the hGH tryptic digest described above. In this experiment, a neutral-loss scan was performed for a loss of 24 mass units, corresponding to the loss of CH_3SH from doubly charged peptides containing methionine (there are three such peptides in the digest). As with the precursor-ion scan, the resulting spectrum is much simpler than the Q1 spectrum (Fig. 2.4A). The largest peak in the spectrum represents peptide T11. A second methionine-containing peptide, T2, is present as a small peak. One expected peptide, T1, is not detected. There are several peaks that do not correspond to methionine-containing peptides but coincidentally lose 24 mass units from other precursors.

Several strengths and weaknesses of neutral-loss scans are revealed by this example. As with precursor-ion scans, efficient fragmentation into the expected product must occur. This was usually the case for one peptide, less often for a second peptide and did not occur with the third peptide. Specificity is again important; the neutral-loss spectrum was greatly simplified, but several abundant ions were formed by losses other than the characteristic CH_3SH. In practice, neutral-loss scans are generally used to identify post-translational modifications of peptides. The loss of phosphoric acid from phosphoserine and phosphothreonine is manifested in positive-ion mode as a neutral loss of 98 mass units (Covey et al. 1991). Unlike precursor-ion scans for small fragments, neutral-loss scans must take into account the charge state of the precursor. A singly charged phosphopeptide can lose 98 mass units; the same peptide, when doubly charged, loses 49 mass units, and so on. Either a particular charge state must be assumed (doubly charged ions are typical of tryptic peptides) or the experiment must be repeated for each possible charge state.

2.4 Ion-Trap MS

2.4.1 Instrument Overview

The arrangement of components in an ion-trap mass spectrometer is, in many ways, similar to that of a triple quadrupole. Ions are produced at atmospheric pressure by an ESI source, transferred through differentially pumped stages, focused into a mass analyser and, finally, directed onto a detector. However, whereas a triple quadrupole performs isolation, collisional activation, and mass analysis in different parts of the instrument (tandem in space), the ion trap performs each operation sequentially in the same volume of space (tandem in time). Much of what makes ion traps unique stems from this distinction. From the structure of the ion-trap mass spectrometer (Fig. 2.2B), it is apparent that ions are formed externally and must be injected – and confined – in the trap itself. This is a significant technical challenge, because ions with an appreciable velocity entering the trap will pass through the trap and continue out the other end (or collide with an electrode and be lost) unless some damping force is applied (March 1998). This is accomplished via the introduction of a helium "bath" gas to the analyser region of the instrument (Stafford et al. 1984; Louris et al. 1989). Ions entering the trap lose their translational energy due to repeated collisions with the bath gas. The cooled ions settle into their predicted oscillating motion near the centre of the trap. The presence of helium (no other inert gas works as well) also dramatically improves the resolution and intensities of peaks in ion-trap mass spectra.

Unlike TSQ mass spectrometers, in which a beam of ions continuously passes through the mass analyser as it is repeatedly scanned, ion traps perform a series of separate operations to acquire each spectrum. Ions are injected from the external electrospray source and are trapped for a given a period of time, allowed to "cool" into stable trajectories and scanned out in a mass-dependent manner. If a product-ion scan is performed, additional steps for isolation and collisional activation are included (Louris and Cooks 1987). Once ions have been trapped, the beam from the electrospray source must be deflected to avoid creating artefactual signals at the detector as the subsequent steps are performed. Since sample is wasted whenever it is unanalysed, ion-trap scans can be characterised by their "duty cycle." If trapping times are long compared with the time spent manipulating and scanning the stored ions, as is often the case for product-ion scans, the duty cycle is quite efficient. However, when trapping times are short, as is typical of full mass-range scans where chemical background ions from the electrospray source are always present, the duty cycle is much less efficient (Korner et al. 1996). Different scan types can require very different ionisation times due to the phenomenon of space charge.

Space charge is an inherent obstacle to using ion traps as practical mass spectrometers. The equations of motion described in Section 2.1 are strictly accurate for only single ions in an electric field. If many ions are present in a confined space, the electric field produced by each ion influences the others, perturbing them from their predicted behaviour; the greater the number of ions forced together, the greater the deviation. This phenomenon is a minor consideration for beam instruments like quadrupoles but is readily observed in ion traps as a loss of resolution and a shifting of peaks to higher apparent masses (Todd and Mather 1980). Under these conditions, accurate calibration of the instrument is almost impossible, because the magnitude of the mass shift depends on how "over-crowded" with ions the trap is during a given scan. This problem has been addressed by an innovation called "automatic gain control" wherein, before each scan is performed, ions are trapped for a brief time and detected without mass analysis. Based on the measured intensity of ions in this "pre-scan", a trapping time is set for the accumulation of ions for mass analysis. In this way, a consistent number of ions is allowed into the trap during every scan.

2.4.2 Full Mass-Range and High-Resolution Scans

Once confined in the trap, ions can be ejected again by processes such as mass-selective instability or resonance ejection. If ejection is combined with scanning of the RF amplitude and an external detector (a conversion-dynode/electron-multiplier apparatus) is engaged, a mass spectrum can be acquired. The mass range scanned is a function of the RF, the radius of the trap and the maximum voltage that can be practically applied. If resonance ejection is employed, the mass range also depends on the q value at which ions are ejected. The Finnigan MAT LCQ, for example, has a normal operating mass range of 2000 mass units, which can be doubled using resonance ejection at a lower q value. Mass accuracy depends on the calibration and stability of the RF generator. Accuracies within 0.1 Da, stable over the course of several weeks, can be obtained if space charging is minimised. The resolution achievable (assuming low space charging) is a function of the rate of scanning; the slower the scan-speed, the better the separation in time between the ejection of ions with adjacent masses. Resolutions of 100,000 ("full width at half the maximum peak height" definition) can be achieved by dramatic reductions in scan speed (Williams et al. 1991). Unfortunately, the influence of space charging also increases as the scan speed drops, so accurate mass assignment becomes more difficult. A compromise is the "zoom scan" implemented on the LCQ, in which a narrow window of masses is scanned with sufficient resolution to distinguish the isotopes of quadruply charged ions. See the inset to Fig. 2.3B for an example of a zoom scan of a doubly charged peptide.

2.4.3 Product-Ion Scans

Once confined in the trap, however, ions need not be scanned out immediately. For instance, they can be isolated, remaining in the trap while those of all other m/z ratios are ejected. Once isolated, an ion can be caused to undergo activation and decomposition. Techniques like photo-dissociation (Louris and Cooks 1987) or surface-induced dissociation (Lammert and Cooks 1991) can be used but, since helium is already present in the same volume as the ions, the easiest approach is to use CID. Resonance excitation can be used to increase the amplitude of motion of a selected ion, causing it to undergo energetic collisions with the bath gas. The process is the same as resonance ejection, but lower voltages are applied, so the precursor stays within the trap and undergoes CID. Following fragmentation, the product ions are scanned out of the trap to produce an MS/MS spectrum (Kaiser et al. 1990; Louris and Cooks 1997).

This experiment is illustrated in Fig. 2.3B. The same hGH tryptic peptide that was analysed by a triple quadrupole was subjected to CID in an ion trap. Comparison with the TSQ spectrum demonstrates some characteristics of MS/MS in the ion trap. As with the TSQ, a complete y-ion series is present, but here the full series of b ions is also seen. The internal fragment Asp–Asn–Ala produced by the quadrupole instrument is absent in the ion-trap spectrum. Both differences result from the way in which CID is performed. The two types of instrument both subject a precursor ion to multiple collisions with an inert gas, but products formed in a TSQ undergo further energetic collisions as they pass through the collision cell and can readily fragment if enough energy is supplied. Therefore, internal fragments are often observed, and b ions, which are less stable than y ions, tend to undergo further decomposition (Tang and Boyd 1992). In the ion trap, activation is supplied by resonance excitation of the precursor ion, so product ions are not activated. An alternative to resonance excitation is broad-band excitation, in which all ions within a defined mass range (as small as a few mass units wide or as large as the entire mass range) are simultaneously activated, resulting in spectra more closely resembling triple-quadrupole CID.

The way in which ion traps perform CID also has other implications. Helium is a much less massive collision gas than argon, so each collision imparts less energy to an ion. Therefore, the internal energy of ions increases in many small steps until it is sufficient to break a chemical bond. Since the internal energy of an ion is distributed among all of its rotational and vibrational modes, the weakest bonds tend to break preferentially. For some molecules, the lowest-energy fragmentation pathway involves the expulsion of water or ammonia, and a few uninformative peaks due to these losses can dominate the ion-trap CID spectra of these compounds. Indeed, such peaks are often the most prominent features of peptide CID spectra, and singly charged peptides tend to yield better product-ion spectra on TSQ instruments. Multiply charged peptides, however, usually produce extensive ion series, as is the case with the hGH tryptic peptide.

A final feature of ion-trap CID is the absence of low-mass ions, including the immonium ions that were so abundant in Fig. 2.3A. This follows from the fact that resonance excitation and resonance ejection are essentially the same phenomenon, differing only in the amplitude to which the ion is excited. Furthermore, the ion is brought into resonance at a particular value of q_z, and its product ions each have their own q_z values, depending on their mass relative to the precursor. In other words, lower-m/z products fall to the right of the precursor, along the $a_z = 0$ line in the stability diagram. If a product ion's q_z value falls outside the stability diagram, it will not be retained in the trap. If the precursor is excited at a low q_z value, more of its products will be trapped than if it is excited at a high q_z value. Resonance excitation more readily ejects ions at low q_z, however, so less energy can be deposited without losing the ion (as opposed to fragmenting it; Louris and Cooks 1987). The default parameters of commercial instruments use a compromise that allows efficient CID while retaining as many of the products as possible. On the LCQ, for example, products with m/z ratios less than approximately one third that of the precursor are lost due to this low-mass limit.

2.4.4 MS^n

Following CID, product ions can be scanned out of the trap to produce a mass spectrum, or they can be manipulated further by additional stages of isolation and CID (MS^n). When, for example, the y_6 ion in the hGH tryptic peptide CID spectrum is isolated and fragmented, the spectrum in Fig. 2.3C is the result. The spectrum resembles that which would be obtained from a truncated version (NH$_2$-DNAMLR-CO$_2$H) of the original peptide. In addition to y ions with the same masses as y ions in the MS/MS spectrum, there are b ions that would have been described as internal fragments had they appeared in the MS/MS spectrum. Recall that, for the full-length peptide, the precursor ion and fragments b_4 and y_4 had the same nominal masses. The MS^3 spectrum resolves this ambiguity. Again, note the low-mass limit; the precursor mass is greater (m/z = 719 vs m/z = 490), so more low-mass ions are lost.

If necessary, more stages of MS can be performed. With each isolation and fragmentation, of course, the number of ions in the trap diminishes. Space-charging considerations limit the number of ions that are stored initially, so there is a practical limit on how far MS^n can be taken. Fortunately, MS^n is most useful in cases where the MS/MS spectrum has only a few prominent ions, so the efficiency of the next isolation is relatively high. As many steps as 11 stages of isolation (MS^{12}) have been performed in extraordinary cases (Louris et al. 1990). Additionally, the spectra generated can become difficult to interpret. As compared with TSQ instruments, CID in ion traps takes place over long time spans, and ions can undergo structural rearrangements. This is particularly so for peptide b ions, and MS^3 on these species can reveal fragments only explicable by cyclisation of the peptide followed by re-opening and subsequent decomposition (Tang et al. 1993; Arnott et al. 1994). MS^n is a powerful tool for structural elucidation, but care must be taken in interpreting the results.

2.5 Virtues of TSQ and Ion-Trap Instruments

Both types of instruments have been used extensively for protein and peptide analysis. The instruments share the ability to perform MS/MS to generate peptide-sequence information that can be used to search protein and DNA databases. ESI has lowered detection limits to such a degree that many proteins can be analysed at biologically relevant levels. The coupling of separation technologies, such as capillary chromatography and electrophoresis, has increased the flexibility of the MS experiments and has further improved sensitivity. When on-line separation is unnecessary, nanospray provides extended analysis times for small volumes of sample, allowing much information to be gathered during a single experiment.

The advantages of one instrument over the other depend largely on the particular experiments to be performed, because a good spectrum is defined as one that answers the question of interest. General figures of merit (such as resolution and sensitivity) can be compared, but not always in a simple way. Both instruments, for example, are capable of full mass-range scanning with detection limits for peptides at low femtomole levels when capillary LC-MS is employed (McCormack et al. 1997; Arnott et al. 1998a). Both can also be operated at unit resolution. For the quadrupole, however, the highest sensitivity is gained at the cost of resolution, so if the instruments are compared when operating at equal resolution, the ion trap may appear to be more sensitive. The ion trap is capable of very high resolution by reducing its scan rate, provided space-charging effects are minimised. Because ion traps can store and accumulate selected ions, sensitivity for MS/MS is greatly enhanced. Product-ion spectra of peptides at the attomole level can be achieved using capillary LC-MS (Arnott et al. 1998b), compared with low-femtomole levels with triple quadrupoles. Sensitivity for MS/MS is likely the greatest advantage possessed by ion traps.

The flexibility of the two types of instruments for diverse experiments can also be contrasted. Ion traps, because they perform stages of MS in the same space, separated by time, cannot readily be used for precursor and neutral-loss scanning. Although some of the same information can be gathered by taking many product-ion spectra over a wide range of precursors, followed by post-acquisition extraction of data, this is a cumbersome exercise and is incompatible with on-line separations. However, ion traps can perform multiple stages of MS with very little effort. This is not often necessary for experiments like protein identification but can be extremely useful for detailed characterisation of post-translational modifications, such as phosphorylation, where the product-ion spectrum is often dominated by neutral loss of the phosphate group. Beam instruments, such as quadrupoles, can only add stages of mass analysis by physically joining more mass spectrometers together. As an alternative, in-source CID can be used to generate fragment ions that can be studied in a TSQ, but this is not true MS^3. Triple quadrupoles have the advantage of being able to perform precursor-ion and neutral-loss experiments. These are used less often than product-ion scans but find important application in the selective detection of modified peptides. No other instrument is as well suited for such experiments. The ion-trap mass spectrometer, because of its sensitivity, flexibility and relatively low cost, is likely to remain a mainstay of proteome research.

References

Arnott D, Kottmeier D, Yates N, Shabanowitz J, Hunt DF (1994) The 42nd ASMS conference on mass spectrometry and allied topics. American Society of Mass Spectrometry, Chicago

Arnott D, Henzel WJ, Stults JT (1998a) Rapid identification of comigrating gel-isolated proteins by ion trap-mass spectrometry. Electrophoresis 19:968–980

Arnott D, O'Connell KL, King KL, Stults JT (1998b) An integrated approach to proteome analysis: identification of proteins associated with cardiac hypertrophy. Anal Biochem 258:1–18

Ballard KD, Gaskell SJ (1993) Dehydration of peptide $(M + H)^+$ ions in the gas phase. J Am Soc Mass Spec 4:477–481

Biemann K (1988) Contributions of mass spectrometry to peptide and protein structure. Biomed Environ Mass Spec 16:1–12

Covey TR, Bonner RF, Shushan BI, Henion J (1988) The determination of protein, oligonucleotide and peptide molecular weights by ion-spray mass spectrometry. Rapid Comm Mass Spec 2:249–255

Covey T, Shushan B, Bonner R, Schroder W, Hucho F (1991) In: Jornvall H, Hoog JO, Gustavsson AM (eds) Methods in protein sequence analysis. Burkhauser Press, Basel, pp 249–256

Davis MT, Stahl DC, Hefta SA, Lee TD (1995) A microscale electrospray interface for on-line, capillary liquid chromatography/tandem mass spectrometry of complex peptide mixtures. Anal Chem 67:4549–4556

Dawson PH (1986) Quadrupole mass analysers: performance, design and some recent applications. Mass Spec Rev 5:1–37

Finnigan RE (1994) Quadrupole mass spectrometers: from development to commericialisation. Anal Chem 66:969A–975A

Griffin L, MacAdoo DJ (1993) The effect of ion size on the rate of dissociation: RRKM calculations of model large polypeptide ions. J Am Soc Mass Spec 4:11–15

Haller I, Mirza UA, Chait BT (1996) Collision induced decomposition of peptides. Choice of collision parameters. J Am Soc Mass Spec 7:677–681

Henzel WJ, Billeci TM, Stults JT, Wong SC, Grimley C, Watanabe C (1993) Identifying proteins from two-dimensional gels by molecular mass searching of peptide fragments in protein sequence databases. Proc Natl Acad Sci USA 90:5011–5015

Huddleston MJ, Bean MF, Carr SA (1993a) Collisional fragmentation of glycopeptides by electrospray ionization LC/MS and LC/MS/MS: methods for selective detection of glycopeptides in protein digests. Anal Chem 65:877–884

Huddleston MJ, Annan RS, Bean MF, Carr SA (1993b) Selective detection of phosphopeptides in complex mixtures by electrospray liquid chromatography mass spectrometry. J Am Soc Mass Spec 4:710–717

Hunt DF, Yates JR III, Shabanowitz J, Winston S, Hauer CR (1986) Protein sequencing by tandem mass spectrometry. Proc Natl Acad Sci USA 83:6233–6237

James P, Quadroni M, Carafoli E, Gonnet G (1993) Protein identification by mass profile fingerprinting. Biochem Biophys Res Commun 195:58–64

Kaiser RE, Cooks RG, Syka JEP, Stafford G (1990) Collisionally activated dissociation of peptides using a quadrupole ion trap mass spectrometer. Rapid Comm Mass Spec 4:30–33

Korner R, Wilm M, Morand K, Schubert M, Mann M (1996) Nano electrospray combined with a quadrupole ion trap for the analysis of peptides and protein digests. J Am Soc Mass Spec 7:150–156

Lammert SA, Cooks RG (1991) Surface induced dissociation of molecular ions in a quadrupole ion trap mass spectrometer. J Am Soc Mass Spec 2:487–491

Louris JN, Cooks RG (1987) Instrumentation, applications and energy deposition in a quadrupole ion trap tandem mass spectrometer. Anal Chem 59:1677–1685

Louris JN, Amy JW, Ridley TY, Cooks RG (1989) Injection of ions into a quadrupole ion trap mass spectrometer. Int J Mass Spec Ion Proc 88:97–111

Louris JN, Brodbelt-Lustig JS, Cooks RG, Glish GL, Berkel GV, McLuckey SA (1990) Ion isolation and sequential stages of mass spectrometry in a quadrupole ion trap mass spectrometer. Int J Mass Spec Ion Proc 96:117–137

Mann M, Wilm M (1994) Error-tolerant identification of peptides in sequence databases by peptide sequence tags. Anal Chem 66:4390–4399

Mann M, Wilm M (1996) Analytical properties of the nanoelectrospray ion source. Anal Chem 68:1–8

Mann M, Hojrup P, Roepstorff P (1993) Use of mass spectrometric molecular weight information to identify proteins in sequence databases. Biol Mass Spec 22:338–345

March RE (1998) Quadrupole ion trap mass spectrometry: Theory, simulation, recent developments and applications. Rapid Comm Mass Spec 12:1543–1554

March RE, Todd JF (eds) (1995) Practical aspects of ion trap mass spectrometry. (Modern mass-spectrometry series, vol 1) CRC, Boca Raton

McCormack AL, Somogyi A, Dongre A, Wysocki V (1993) Fragmentation of protonated peptides: surface-induced dissociation in conjunction with a quantum mechanical approach. Anal Chem 65:2859–2872

McCormack AL, Schieltz DM, Goode B, Yang S, Barnes G, Drubin D, Yates JR III (1997) Direct analysis and identification of proteins in mixtures by LC/MS/MS and database searching at the low-femtomole level. Anal Chem 69:767–776

Roepstorff P, Fohlman J (1984) Proposal for a common nomenclature for sequence ions in mass spectra of peptides. Biomed Mass Spec 1:601

Smith RD, Loo JA, Edmonds CG, Barinaga CJ, Udseth HR (1990) New developments in biochemical mass spectrometry: electrospray ionization. Anal Chem 62:882–899

Stafford, GC, Kelley PE, Syka JE, Reynolds WE, Todd JF (1984) Recent improvements in, and analytical applications of advanced ion trap technology. Int J Mass Spec Ion Proc 60:85–98

Tang X-J, Boyd RK (1992) An investigation of fragmentation mechanisms of doubly protonated tryptic peptides. Rapid Comm Mass Spec 6:651–657

Tang X-J, Thibault P, Boyd RK (1993) Fragmentation reactions of multiply-protonated peptides and implications for sequencing by tandem mass spectrometry with low-energy collision-induced dissociation. Anal Chem 65:2824–2834

Todd JFJ, Mather RE (1980) The quadrupole ion store (Quistor) part IX: Space-charge and ion stability. A theoretical background and experimental results. Int J Mass Spec Ion Proc 34:325

Wahl JH, Gale DC, Smith RD (1994) Sheathless capillary electrophoresis-electrospray ionization mass spectrometry using 10 µm I.D. capillaries: analyses of tryptic digests of cytochrome c. J Chrom A 659:217–222

Watson JT (1997) Introduction to mass spectrometry. Lippincott-River, Philadelphia

Whitehouse CM, Dreyer RN, Yamashita M, Fenn JB (1985) Electrospray interface for liquid chromatographs and mass spectrometers. Anal Chem 57:675

Williams JD, Cox K, Cooks G, Schwartz JC (1991) High mass resolution using a quadrupole ion trap mass spectrometer. Rapid Comm Mass Spec 5:327–329

Wilm MS, Mann M (1994) Electrospray and Taylor-Cone theory, Dole's beam of macromolecules at last. Int J Mass Spec Ion Proc 136:167–180

Wilm M, Neubauer G, Mann M (1996a) Parent ion scans of unseparated peptide mixtures. Anal Chem 68:527–533

Wilm M, Shevchenko A, Houthaeve T, Breit S, Schweigerer L, Fotsis T, Mann M (1996b) Femtomole sequencing of proteins from polyacrylamide gels by nano-electrospray mass spectrometry. Nature 379:466–469

Yates JR III, Speicher S, Griffin PR, Hunkapiller T (1993) Peptide mass maps: a highly informative approach to protein identification. Anal Biochem 214:397–408

Yates JR III, Eng JK, McCormack AL (1995) Mining genomes: correlating tandem mass spectra of modified and unmodified peptides to sequences in nucleotide databases. Anal Chem 67:3202–3210

3 The Basics of Matrix-Assisted Laser Desorption, Ionisation Time-of-Flight Mass Spectrometry and Post-Source Decay Analysis

BERNHARD SPENGLER

3.1 Introduction

Mass spectrometry (MS) has changed its appearance in the scientific world considerably during recent years. MS started out as a tool in atomic physics approximately 100 years ago and has gradually been applied to the life sciences. Less than 20 years ago, a new ionisation method, fast atom bombardment, made MS a viable tool in the field of biochemistry (Barber et al. 1981). Approximately 10 years ago, with the invention of electrospray ionisation (ESI; Fenn et al. 1989) and matrix-assisted laser desorption ionisation (MALDI; Karas and Hillenkamp 1988), MS became a major tool for biology, biomedicine and molecular medicine. The impressive progress of MS can be summarised by plotting two methodological parameters of MS, accessible mass range and sensitivity, against one another (Fig. 3.1). The dramatic extension of mass range and sensitivity due to the development of ESI and MALDI were two of the essential prerequisites for the use of MS in the field of modern biology.

Among the most attractive properties of MS in the field of bioanalysis are:

- The high informational content of MS data
- The high sensitivity
- The rapidity of data accumulation
- The versatility
- The accuracy of the method

Today, MALDI has successfully entered the field of medical research (for example, in cellular diagnostics, gene therapy and cancer research). Investigation of complex (multi-component) samples and the direct characterisation of cellular material are two typical applications of MALDI. In proteome research, MALDI is routinely employed for protein characterisation, typically after two-dimensional (2D) gel electrophoresis and enzymatic digestion. Although MALDI is limited to providing molecular-weight information only (rather than structural information), its spread in the proteome field is unprecedented. Another method, MALDI post-source decay (PSD; Kaufmann et al. 1993) has further increased the acceptance of MALDI MS by providing additional structural information from the same samples with the same principal parameters as for MALDI.

Mass Spectrometry

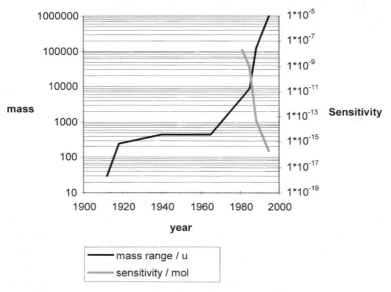

Fig. 3.1. Progress of mass-spectrometric techniques with time, expressed as the two parameters "accessible mass range" and "analytical sensitivity"

PSD in combination with MALDI analysis allows one to perform highly sensitive characterisation of peptides and proteins directly from complex samples or from native biological fluids or tissues. A pertinent example is the direct sequencing of neuropeptides from the pituitary pars intermedia of the amphibian *Xenopus laevis*. Figure 3.2 shows the MALDI spectrum obtained directly from the cell extract of the biological tissue. Immediately after recording this spectrum, several of the peptide signals were further investigated by PSD analysis, and the corresponding amino acid sequences could be determined (Jespersen et al. 1999).

Major histocompatibility complex (MHC) peptide analysis is another example of the usefulness of biomedical analysis by MALDI MS. Combining affinity chromatography, high-performance liquid chromatography (HPLC) fractionation and MALDI-PSD analysis, a peptide pool of approximately 10,000 components harvested from human renal-carcinoma cell lines was investigated and some of the components were identified as cancer-specific peptides (Flad et al. 1998). In this and many other projects, MALDI MS is directly involved in biomedical research, proteomics or biotechnology. The aim of this chapter is to summarise the basic principles and the methodology of MALDI and MALDI-PSD. Applications of these techniques will be referred to in some of the following chapters.

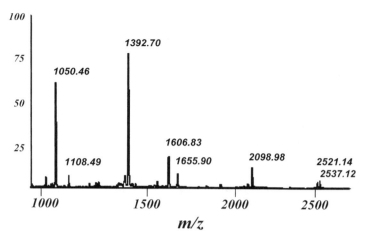

Fig. 3.2. Matrix-assisted laser desorption ionisation mass spectrum of pars intermedia tissue from the amphibian *Xenopus laevis*, displaying several previously known neuropeptides and some unexpected neuropeptides. (Jespersen et al. 1999)

3.2 Principles of MALDI

MALDI was developed as an extension of the older technology of laser desorption ionisation (LDI), which had been used extensively (and still is being used) in the field of microprobe analysis (Van Vaeck and Gijbels 1990a, 1990b). LDI of soluble compounds is based on the simple approach of air-drying analyte solutions on a metal target and forming ions [using an ultra-violet (UV) laser pulse] that can be detected by time-of-flight (TOF) mass analysis. The main limitation of this approach is that analyte substances usually have rather different spectral absorptions at the laser wavelength used and often are completely transparent. Only highly absorbing molecules are accessible by LDI, while non-absorbing molecules can only be ionised at the expense of extensive fragmentation. The useful mass range is rather low (m < 1000 u), because the thermal stress from absorption of the laser light does not allow the desorption of larger molecules as intact entities. Furthermore, even under ideal conditions, LDI has a rather low ionisation yield (and, thus, analytical sensitivity) and mainly produces neutral desorbed material (Spengler et al. 1988).

All of these disadvantages are circumvented by detaching the energetic processes necessary for desorption and ionisation from the analyte molecules; this is done by employing an intermediate matrix for energy transfer. The feasibility of this approach was first demonstrated when mixtures of absorbing and non-absorbing substances were investigated. Although the non-absorbing amino acid alanine (as a neat sample) was not accessible by LDI, it gave clear MS signals when mixed with the highly UV-absorbing amino acid tryptophan

(Karas et al. 1985). Tryptophan was acting here as a *matrix* for the analyte alanine by absorbing the energy of the laser pulse; it acted as an intermediary for the desorption and ionisation of *both* components. This observation led to the development of the technique MALDI by using a highly UV-absorbing material considerably in excess of the analyte under investigation (Karas et al. 1987).

3.2.1 Ion Formation by MALDI

The mechanisms of ion formation from solid material by laser irradiation were found to consist of at least three different pathways; each contributes to the ionisation of molecules to a variable degree, depending on the experimental conditions (Spengler et al. 1987). The pathways are:

1. Photo-ionisation by individual photon–molecule interactions, as in gas-phase photo-ionisation of single molecules
2. Ionisation by protonation or de-protonation via interactions between excited, ionised and/or protonated (or de-protonated) molecules and neutral molecules, as in chemical ionisation mechanisms
3. Cluster decay of small particles formed by lattice disintegration, leading to charge inhomogeneities and charging of the remaining molecular entities, as in liquid-spray methods

The introduction of the MALDI method has overcome the strong substance-specific variability that exists among the various ion-formation processes. The principle of MALDI is shown in Fig. 3.3. Desorption of intact individual molecules from neat samples (left side of the figure) becomes a dominant problem as the mass of the analyte increases, because the *inter*molecular forces between biomolecules tend to approach the range of *intra*molecular forces. This means that, if the accumulated internal energy is sufficient for desorption of material, it is also sufficient to cause considerable fragmentation. However, the matrix in the MALDI approach (right side of the figure) allows the separation of the analyte molecules from each other and allows only small interactions between the analyte molecules and the target. Matrix substances are usually relatively volatile in comparison to the analyte molecules, leading to an efficient desorption ("evaporation") of material.

Under MALDI conditions, it was found that pathway (2) in the upper scheme of Fig. 3.3 plays the most important role. This was demonstrated by several observations, including direct imaging of the desorption plume (Bökelmann et al. 1995). Using a small TOF analyser that could be tilted around the sample in a field-free vacuum housing, it was possible to detect the angular and velocity distributions of ions desorbed by pulsed laser irradiation. Figure 3.4 summarises the observations for a typical MALDI sample of 2,5-dihydroxybenzoic acid (as a matrix) and substance P (as the analyte), mixed in a molar ratio of 1000 : 1. It shows the rather broad distribution of velocities and angles the TOF

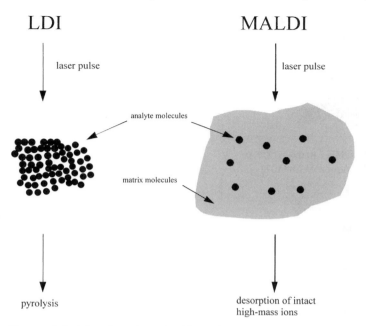

Fig. 3.3. Principles of matrix-assisted laser desorption ionisation

analyser has to cope with when attempting high mass-spectral resolution. It also shows that (protonated) analyte ions are present in an area where (protonated) matrix ions are diminished. Thus, analyte ions are obviously formed in the gas phase by proton transfer from protonated matrix molecules to neutral analyte molecules. This "overlap model" (Bökelmann et al. 1995) provides a basis for an understanding of several phenomena in MALDI-TOF MS, including PSD and delayed extraction (DE).

The formation of protonated matrix ions can be understood as a primary event in MALDI; it is probably induced by the initial formation of radical ions (Spengler et al. 1987; Ehring et al. 1992). The most plausible causative pathway is the photo-ionisation of highly excited matrix molecules in the condensed phase or in a dense area of the gas phase, followed by interaction with other matrix molecules to form protonated matrix molecules and neutral matrix radicals:

$$M^{**} \rightarrow M^{\bullet +} + e^- \tag{1}$$

$$M^{\bullet +} + M \rightarrow [M+H]^+ + [M-H]^\bullet \tag{2}$$

The formation of radical ions is probably not a prerequisite in *infrared MALDI*, because the considerably lower photon energies used in this technique would not favour photo-ionisation. The basic properties of a "good" MALDI matrix are:

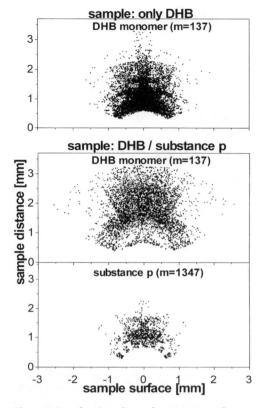

Fig. 3.4. Ion-density plots of matrix ions from a neat matrix sample (*top*), of matrix ions from a regular matrix-assisted laser desorption ionisation (MALDI) sample prepared from 2,5-dihydroxybenzoic acid (as the matrix) and substance P (as the analyte; *middle*), and of substance-P ions from the same MALDI sample (*bottom*). The density plots depict the spatial distribution of desorbed ions above the sample surface at an arbitrarily chosen observation time of 1 μs after the laser pulse. The images show that matrix ions are lacking in areas of high analyte-ion densities

- A degree of volatility allowing both vacuum stability and efficient desorption
- A high spectral absorption at the laser wavelength used
- Good mixing and solvent compatibility with the analyte
- An appropriate level of proton affinity to allow for protonation or deprotonation ("ionisation") of the co-desorbed analyte molecules

Although the mechanisms of ion formation have been extensively investigated, useful matrices have nevertheless been found empirically. Some of the most common matrices used in MALDI are listed in Table 3.1. The common features of these "working" matrices are (in addition to the properties mentioned earlier) that they are easily prepared and have a high tolerance towards common contaminants and sample additives.

Table 3.1. Common matrix substances used for matrix-assisted laser desorption ionisation mass spectrometry

Matrix	Wavelength	Comments	Reference
DHB	337 nm, 355 nm		Strupat et al. (1991)
α-Cyano-4-hydroxycinnamic acid	337 nm, 355 nm	Useful for thin-film "add-on" preparation	Beavis et al. (1992)
Sinapinic acid	337 nm, 355 nm		Beavis and Chait (1989)
3-Hydroxypicolinic acid	337 nm, 355 nm	Useful for oligonucleotides	Wu et al. (1993)
DHB plus 10% 2-hydroxy-5-methoxybenzoic acid	337 nm, 355 nm	Useful for large molecules weighing more than 20 kDa	Karas et al. (1991)
Nicotinic acid	266 nm		Karas and Hillenkamp (1988)
Succinic acid	2.94 μm, 10.6 μm		Nordhoff et al. (1992)
Glycerol	2.94 μm, 10.6 μm		Overberg et al. (1990)

DHB, 2,5-Dihydroxybenzoic acid

3.2.2 Sample Preparation

Two different techniques of sample preparation are commonly used: "dried-droplet" and "add-on" or "surface" preparation. In the *dried-droplet technique*, solutions of matrix and analyte are mixed directly on the target or in an Eppendorf cup. The solvents used for the two solutions are usually identical (or at least compatible) and can be pure water or mixtures of water with organic solvents (acetonitrile, acetone or ethanol). The matrix solution is typically saturated; i.e. the concentrations are in the range of 5–50 g/l, depending on the matrix substance and the solvent. The concentration of the analyte solution can be in the range of 10^{-4} M to 10^{-7} M. Proteins and peptides are often prepared with additional trifluoroacetic acid (0.1%). Approximately 0.5 μl of the matrix–analyte mixture are pipetted onto the metal target and are dried before introduction into the mass spectrometer. Drying can be done either slowly (at room temperature) or quickly (on a preheated target or with a stream of hot air). The former procedure leads to the growth of large matrix crystals and to a strong degree of analyte incorporation into the crystals. The effect is a purification of the analyte (because, e.g. salts are not incorporated into the matrix crystals); another effect, however, is a rather high sample inhomogeneity across the preparation spot. The latter procedure, however, causes the formation of rather homogeneous, small crystals but lacks the purification effect (and, thus, the improved tolerance of contaminants).

The *add-on* or *surface technique* (Vorm and Mann 1994) uses incompatible solvents in an approach directly opposed to the dried-droplet technique, in which one tries to avoid mixing of the matrix and analyte. In the first step of the procedure, a thin, homogeneous layer of micro-crystalline matrix is formed by rapid evaporation of matrix material dissolved in a volatile solvent (acetone). The technique is applicable only if the matrix is weakly soluble in water, e.g. α-cyano-4-hydroxycinnamic acid. Before adding the analyte, the matrix layer can be washed with water to remove salt impurities. This is done by applying a few microlitres of cold water to the matrix layer for several seconds before the solvent is blown from the surface. In the second step, $0.5\,\mu l$ of aqueous analyte solution is added to the surface and air-dried. Another washing step, which often results in an improved signal-to-noise ratio, can be added (Vorm and Mann 1994). The advantage of the surface technique is the formation of very homogeneous thin samples, which provide increased sensitivity in MALDI-MS due to the concentration of the analyte in the uppermost accessible layers of the sample. The sensitivity of MALDI has been improved so that sample amounts necessary for analysis are in the range of a few femtomoles (dried-droplet technique) or even attomoles (add-on technique).

3.2.3 Instrumentation

3.2.3.1 Ion Source

The most common type of ion source uses a pulsed nitrogen laser at $337\,nm$ for desorption ionisation. This is a balance between the UV absorption of the most useful matrices and the wavelength-dependent ionisation behaviour (it is also used because commercial nitrogen lasers are very user friendly). Laser wavelengths that are less often used include $355\,nm$ and $266\,nm$ (from the frequency-tripled and frequency-quadrupled neodymium:yttrium-aluminium garnet laser), $308\,nm$ (from excimer lasers), approximately $220\text{--}350\,nm$ (from dye lasers), $2.94\,\mu m$ (from erbium:yttrium-aluminium garnet lasers) and $10.6\,\mu m$ (from carbon dioxide lasers). The irradiance of the laser beam on the target has to be carefully adjusted, since both very low and high irradiances produce unusable spectra. Ion acceleration is usually achieved with a single or dual electrostatic acceleration field [although, recently, electrodynamic (DE) fields have been used, as discussed later].

3.2.3.2 Mass Analyser

MALDI is typically coupled to TOF mass analysers due to the pulsed nature of both techniques. It has also been coupled to other mass analysers, such as:

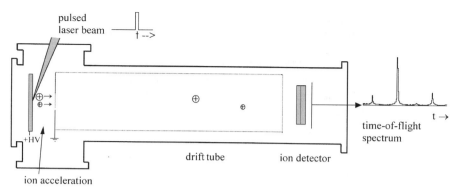

Fig. 3.5. Scheme of a simple linear time-of-flight mass spectrometer with a pulsed laser source

Fourier-transform ion cyclotron-resonance instruments (Castoro and Wilkins 1993; Li and McIver 1994), sector-field instruments (Annan et al. 1992) and ion traps (Doroshenko and Cotter 1995; Qin and Chait 1996). The principle of MALDI TOF mass analysis is shown schematically in Fig. 3.5. The starting event for the flight-time measurement of ions is defined by the laser pulse. Ions are accelerated by a strong electric field in the range of 5–30 kV over a distance of 2–20 mm. After acceleration, ions enter a field-free drift region between 0.5 m and 3 m long. The mass-dependent flight times of the mono-energetic ions through the drift region are used for mass analysis. In addition to the linear TOF instruments shown in Fig. 3.5, reflectron instruments are used to compensate the initial energy spreads and improve the mass resolution and mass accuracy (Karas and Hillenkamp 1988).

3.2.3.3 Ion Detector

Ions are usually detected by microchannel plate (MCP) detectors or by secondary electron multipliers (SEM). The impinging ions desorb electrons from the detector surface; these electrons are then multiplied in cascades due to acceleration through the channels or by acceleration onto the subsequent dynodes, respectively. The flat surface of MCPs provides for retention of a high flight-time resolution (and thus mass resolution). Problems occur with MCP detectors for higher masses of ions. The detection efficiencies of these detectors are known to be velocity dependent, with a threshold velocity in the range of $2 \cdot 10^4$ m/s Geno and Macfarlane 1989. Since the velocity of accelerated ions decreases according to one over the square root of the mass, ion detection becomes inefficient for analyte ions of high molecular weight. Post-accelerating detectors, employing a first "conversion dynode" in front of a regular SEM or MCP, are used especially for the high-mass range. The high accelerating potential of the conversion dynode causes an increased velocity of ions prior to detection. The most important effect of such post-accelerating detectors is that, in addition to electrons, small secondary ions are desorbed from the surface of the conversion dynode; these ions are then accelerated onto the subsequent SEM or MCP and are detected with high efficiency (Spengler et al. 1990). The detection efficiency of large ions is dramatically increased by this technology, but the high mass resolution is sacrificed. The fast analogue signal from the detector is digitised and stored by a digital oscilloscope or a transient recorder and is processed by a computer.

3.2.3.4 Delayed Ion Extraction

The performance of MALDI TOF MS has dramatically improved due to the introduction of the DE technique. The method has been found to be very useful in enhancing the flight-time focussing of ions formed by MALDI and results in a large increase in mass-resolving power (Colby et al. 1994; Brown and Lennon 1995; Vestal et al. 1995; Whittal and Li 1995). The principle of DE is rather old and was used in the early days of TOF mass analysis (Wiley and McLaren 1955). For MALDI MS, it was used in an orthogonal configuration in one of the very early MALDI instruments (Spengler and Cotter 1990), though with limited success in improving resolution. DE is based on the concept that ions of different initial velocities can be focussed according to flight time by using appropriately pulsed acceleration fields. Electrostatic fields do not allow ions of the same mass to be focussed to a certain flight time if these ions already have a considerable initial velocity spread prior to acceleration. Ion reflectors, for example, are able to compensate for flight-time distributions due to "pure energy errors", i.e. energy distributions that ions have when they pass a certain point in the field-free drift region at the same time (a time focus). They are, however, unable to fully compensate flight-time errors resulting from energy distributions of ions prior to acceleration, because these kinds of energy distributions correspond to a combination of pure energy errors (which can be compensated for) and start-time errors (which can never be compensated for via static methods).

For MALDI ions, it is known that the initial velocity distribution is rather broad (Fig. 3.4) and not strongly dependent on mass. Using the correlation between the initial velocity of an ion formed by MALDI and its position in a field-free region above the sample surface after a certain flight time, it is possible to compensate for these timing errors. If the acceleration field is turned on at a certain delay time after the laser pulse, ions with a high initial velocity (which would usually reach the detector too early) will start with a lower acceleration potential and will thus gather less kinetic energy by acceleration than ions with a low initial velocity. With an optimised combination of delay time and pulse voltages in a two-stage acceleration system, flight-time errors of MALDI ions can be compensated for almost completely. The DE method can be optimised for both linear and reflectron TOF instruments.

A mass-resolving power in the range of 4000 can be reached for MALDI ions in a linear instrument, whilst reflectron instruments can have a resolving power of up to 10000 or even 15000 (Vestal et al. 1995). The consequences of the DE technique for PSD analysis will be discussed later.

3.3 Principles of MALDI-PSD Mass Analysis

In an attempt to investigate the reasons for the reduced mass-resolving power of MALDI instruments in the early days of the technique, it was found that ions formed by MALDI are much less stable during flight than was expected (Spengler et al. 1991). The observation that very large molecules could be desorbed, ionised and detected by MALDI had led to the assumption that MALDI is a very "soft" ionisation method with very little internal energy uptake. In reality, it was found that large ions formed by MALDI are rather unstable and that the decay of these ions takes longer than that of small ions. As a result of the longer decay times, fragment ions are typically not observed in linear instruments, because fragmentation takes place after acceleration, in the field-free drift region of the instrument. Therefore, the flight times of the products are unchanged and, hence, the detected mass appears to be that of the intact ion. In reflectron instruments, the product ions are detectable if the transmission and detection characteristics are optimised for these types of ions, since they are decelerated and re-accelerated.

3.3.1 Mechanisms for Ion Activation

Ion fragmentation is the result of internal energy gathered by electronic, vibrational or rotational excitation of the molecular ions during the desorption or ionisation processes, or afterwards, through other processes. In MALDI MS, it was found that at least three different pathways are responsible for the transfer of internal energy to the molecular ions. These are summarised in Table 3.2.

Post-source activation (mode 3) was investigated first, as a function of the residual-gas pressure of the instrument (Spengler et al. 1992). Figure 3.6 shows the decrease of ion stability with increasing residual-gas pressure. At a residual-gas pressure of, e.g. $2 \cdot 10^{-7}$ hPa after passing through a flight tube of 3 m, only 78% of cytochrome C ions reach the detector as intact molecular ions; the remaining 22% are fragmented.

Depending on the vacuum conditions of the instruments, fragmentation of ions can result from high-energy collisions with residual-gas molecules to a considerable extent. As can be seen from the example in Fig. 3.6, no intact ions of cytochrome C are obtained with a residual-gas pressure of of 10^{-5} hPa, and only approximately 20% of the ions reach the detector intact at a residual-

Table 3.2. Activation mechanisms in matrix-assisted laser desorption ionisation that contribute to formation of post-source decay ions

Mode	1	2	3
Activation mechanism	Sample activation	In-source activation	Post-source activation
Activation region	Surface/sample	Selvedge	Vacuum
Activation processes	Direct photon-molecule interactions, solid state activation, temperature effects	Multiple low-to-medium-energy collisions	High-energy collisions
Time scale	Picoseconds to nanoseconds	Nanoseconds to microseconds	Microseconds to milliseconds
Distance from surface	0	<200 µm	Farther than the acceleration zone

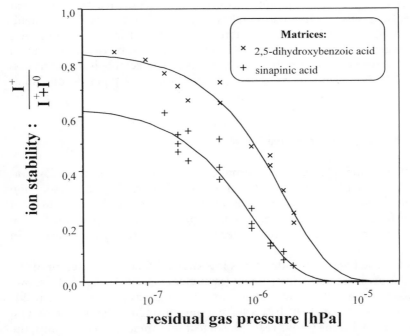

Fig. 3.6. Ion stability of cytochrome C (molecular weight = 12,360 u) as a function of the residual-gas pressure for 2,5-dihydroxybenzoic acid (DHB), and sinapinic acid (as a matrix). Ion-stability values are determined for a drift path of 3 m. The resulting reaction cross sections of cytochrome C with 20 keV kinetic energy are 7.0 nm² for DHB and 14.1 nm² for sinapinic acid

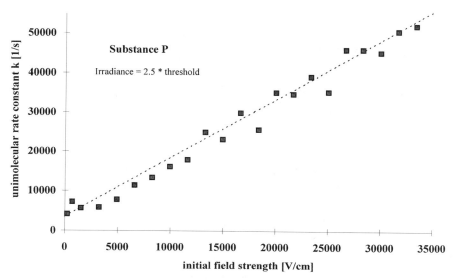

Fig. 3.7. Field dependence of the unimolecular rate constant k of substance P at zero (extrapolated) residual-gas pressure

gas pressure of $1 \cdot 10^{-6}$ hPa with sinapinic acid as a matrix. Collision theory allows the calculation of the reaction cross-section σ for fragmentation due to high-energy collisions:

$$\sigma = -\frac{kT}{L \cdot P} \ln \frac{I^+}{I^+ + I^0} \tag{3}$$

where L is the flight path length, p is the residual-gas pressure, T is the absolute temperature, I^+ is the signal intensity of stable ions and I^0 is the signal intensity of neutral fragmentation products.

Even when such curves are extrapolated to zero residual-gas pressure, a considerable degree of ion fragmentation can still be observed. The activation pathway for this fragmentation is listed as mode 2 (in-source activation) in Table 3.2. Mode-2 activation was found to be strongly dependent on the electrical-field strength in the first acceleration region. Figure 3.7 shows the dependence of extrapolated ion stabilities (for zero residual-gas pressure) at various initial field strengths.

The mechanism of this type of activation is based on multiple low- to medium-energy collisions between analyte ions and neutral matrix molecules within a distance of a few hundred micrometers from the sample surface during acceleration by the extraction field (Bökelmann et al. 1995). Figure 3.7 shows that, even at zero initial field (and at zero residual-gas pressure), there is still a considerable degree of fragmentation. Ion activation for this type of decay is listed as mode 1 ("sample activation") in Table 3.2. It can be interpreted as the result of various surface processes, such as direct photon–molecule interactions, solid-state activation or temperature effects.

It has to be noted that all of these activation mechanisms lead to fragmentation only in the field-free drift tube of the instrument, even if the ions were activated on the surface or in the acceleration region. The name "post-source decay" was introduced to summarise these three pathways and is not, therefore, a mechanistic description; it simply describes all the kinds of fragmentation events that take place in the field-free drift region in MALDI MS.

In MALDI instruments with prompt ion extraction (without DE) and under good vacuum conditions, mode 2 is the most important contributor to ion activation. Modern MALDI-PSD instruments, however, usually employ DE. Since the density of the cloud of neutral matrix molecules is already reduced to some extent when the extraction field is turned on (typically a few hundred nanoseconds after the laser pulse), in-source activation of ions can be assumed to be considerably reduced. In fact, the degree of PSD is considerably reduced under DE conditions, but the improvement in the signal-to-noise ratio of PSD ion signals due to enhanced mass resolution mostly compensates for this effect (Kaufmann et al. 1996). In cases where the degree of fragmentation is too low, PSD can be increased by enhancing mode 3 of ion activation, i.e. by increasing the residual-gas pressure of the instrument or by employing a collision cell, which is an option with some of the commercial instruments.

3.3.2 Instrumentation for MALDI-PSD

Instrumentation for MALDI-PSD is similar to that of MALDI instrumentation. A few things, however, are important (or even essential) for PSD analysis. These are summarised below.

3.3.2.1 Ion Source

Instrumental ion transmission is, to a considerable extent, a function of the geometry and electrical potentials of the ion source. This is true for both MALDI-PSD and for MALDI, because PSD takes place mainly after leaving the ion source. To avoid a considerable loss of PSD ions that are formed early during ion acceleration, the ion-source region should be as short as possible. This allows the precursor ions to become fully accelerated prior to decay. Increasing the electric field during the first acceleration stage helps to increase in-source activation but, as mentioned earlier, the effect is less pronounced with DE ion sources.

3.3.2.2 Ion Gate

The selection of a precursor ion from a multi-component MALDI spectrum is one of the most important features of MALDI-PSD instruments. This low-resolution MS-MS capability allows one to perform peptide sequencing directly from the tryptic digestion of a protein (in gel or in solution) without the need for chromatographic separation. The instrumental unit employed for this selection step is usually called an ion gate and is an electrical deflection device. If the device is enabled, all ions are deflected off the detector-beam path except during a certain flight-time window. To achieve this, the deflection plates, wires or strips are fed with a permanent potential of a few hundred volts, which is changed to zero volts for a very short period by a very fast pulse. This shutdown pulse allows the ion package of interest to be transmitted through the instrument to the detector without deflection. Since the ion gate is located before the ion reflector, the same gating pulse transmits precursor ions together with their PSD product ions.

The performance of the ion gate (transmission and selectivity) is dependent on its electrical and geometrical properties. Since a considerable portion of the low-energy (low-mass) PSD ions is formed prior to passing the ion gate, any remaining electrical field has to be avoided, since even a weak field could easily deflect these sensitive ions. The geometry and position of the ion gate determines the accessible selectivity of the gate. An $M/\Delta M$ value of 200 can typically be reached, and improvements are under way.

3.3.2.3 Mass Analyser

The ion reflector of the mass analyser is necessary for the flight-time dispersion of PSD ions, since the relative masses of these ions are expressed not in their velocities (as is the case for stable ions) but in their kinetic energies. Since all PSD ions of a certain precursor ion have the same velocity, the ion reflector is used to give these ions different flight paths (and thus flight times) by acting as an energy analyser.

There is no ideal instrumental solution for performing this task. The potentials of regular ion reflectors have to be stepped down in order to be able to focus the complete mass range of PSD ions onto the detector with sufficient mass resolution. Various systems, such as gridless (inhomogeneous-field) reflectors and single- and dual-stage, gridded (homogeneous-field) reflectors, have been employed. Other solutions, such as the curved-field reflector (Cordero et al. 1995), can be used without the need to step down the potentials, but they cause a rather low ion transmission due to divergent electrical fields and non-ideal geometry constraints.

3.3.3 Interpretation of PSD Mass Spectra

Computerised acquisition control and data evaluation are central tenets of PSD MS. Both instrumental control and the interpretation of spectra are highly demanding tasks compared with the MALDI analysis of stable ions. Stepping down the reflector potentials and the exact calibration of PSD ion masses require highly sophisticated software and precise knowledge of the instrumental parameters.

Spectral interpretation for peptide and protein analysis is based on fundamental investigations of fragmentation behaviour and fragmentation mechanisms and on accumulated PSD data from known peptide sequences. Figure 3.8 shows an example of a PSD product-ion mass spectrum of a peptide. The spectral parts acquired for the various reflector potentials are combined, and a mass-calibration function is applied to the spectrum. Ion signals interpreted as corresponding to N- or C-terminal fragments are labelled using the standard nomenclature (Roepstorff and Fohlmann 1984; Johnson et al. 1988). The determined sequence is displayed in the upper part of the figure. *Dots* at the tops of peaks indicate the presence of certain characteristic neighbouring signals.

The interpretation of a spectrum like that shown in Fig. 3.8 without any prior information regarding the investigated peptide is a rather complicated task. This is because the rules for the fragmentation of peptides are quite complex, and they do not always lead to unequivocal spectral information. In addition to the formation of various types of fragment ions that contain the N-terminus or the C-terminus, losses of H_2O or NH_3 are very common. Furthermore, internal ions, which contain neither end of the peptide, are observed with rather high intensities in PSD analysis. Internal ions are the result of at least two consecutive fragmentation steps. They are more common in PSD MS compared with other techniques, due to the long effective decay time available for fragment-ion formation. These long decay times (between the time the sample leaves the ion source and the time it enters the ion reflector) are the basis of the high sensitivity of the method, because fragment-ion formation becomes very efficient. The decay times are also the reason for the observed high signal intensities of internal ions from secondary or higher-order processes. Internal ions certainly contain additional structural information that can be helpful for spectral interpretation, but they can make spectral information ambiguous. Manual interpretation of fragmentation spectra is discussed in detail in Chapter 8. Software tools help to interpret PSD spectra, and there are strategies available to automate data evaluation for peptide sequencing by MALDI-PSD (Chaurand et al. 1999).

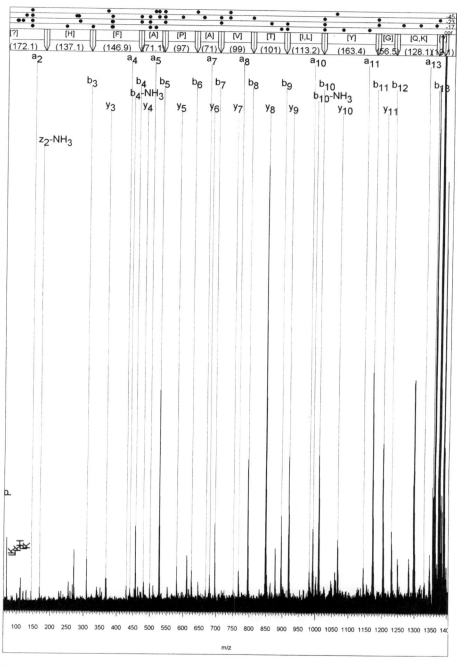

Fig. 3.8. Matrix-assisted laser desorption ionisation post-source-decay mass spectrum of the peptide GDHFAPAVTLYGK; [M + H]⁺ = 1375.8 u. Only ion signals found to be due to N-terminal or C-terminal ions are labelled. The sequence determined by the interpretation software is displayed in the upper part of the figure. *Dots* at the top of the figure signal the presence of characteristic neighbouring signals

3.3.4 Improving Spectral Information

As a consequence of the ambiguity of information observed in certain cases of the PSD analysis of peptides, various strategies for improving the spectral information have been developed. Chemical derivatisation methods – including hydrogen-deuterium exchange (Spengler et al. 1993), N-terminal acetylation (Chaurand et al. 1999), charge tagging (Liao et al. 1997; Spengler et al. 1997) and arginine blocking (Spengler et al. 1997) – are used to simplify the spectra. These methods are not always applicable, however, because they usually cause a decrease in sensitivity and sometimes reduce spectral quality. Hydrogen–deuterium exchange is certainly the easiest derivatisation procedure, and it provides a high degree of additional structural information. Using a simple procedure (Spengler et al. 1993), an almost complete exchange of hydrogen with deuterium is possible, allowing one to determine the number of exchangeable protons, i.e. protons attached to heteroatoms. Most isobaric peptides or peptide fragments can be distinguished with this approach, and peptide-sequence analysis can be made much easier by comparing the PSD spectra of the native and deuterium-exchanged peptides.

Charge tagging, eventually in combination with arginine blocking, allows a degree of control over the formation of fragment ions, because the location of the charge within the peptide chain is fixed and known. Especially in cases of complex fragment-ion formation (with peptides having a basic amino acid in the middle of the chain), charge tagging can considerably simplify PSD. The procedure, however, requires more material than is necessary for a normal PSD analysis and is relatively time consuming.

3.4 Conclusion

MALDI and PSD analysis have become valuable tools for biomolecule detection and characterisation. The principle differences between the MALDI TOF approach and other techniques, such as ESI triple-quadrupole MS or ion-trap MS, are listed below:

1. Disadvantages of MALDI-PSD:
   - Instrumental limitations
   - Complex spectra are difficult to interpret
   - Fewer interfaces have been developed
2. Advantages of MALDI-PSD:
   - High sensitivity
   - Universality
   - Complexity of samples
   - Compatibility with routine MALDI applications (digest analysis)

3. Promising aspects of MALDI-PSD:
   Coupling to surface techniques, such as:
   - 2D separation
   - Offline coupling to capillary electrophoresis and micro-HPLC
   - Biological applications
   - Immobilization techniques (functionalised surfaces)

Due to the solid-phase nature of the MALDI technique, on-line coupling to liquid-separation techniques is not straightforward. Automated off-line coupling, however, can be of interest, especially if sophisticated surface techniques are involved, such as protein immobilisation, surface functionalisation, 2D separation or biological micro-probing. Especially in the field of proteome research, these strategies will make MALDI and MALDI-PSD essential tools in the future. Data evaluation for peptide-sequence analysis by MALDI-PSD is fairly advanced, allowing amino-acid-sequence determination of unknown peptides. This can be performed for rather complex mixtures, such as protein digests or MHC-peptide pools (Flad et al. 1998). Intelligent database searching in combination with automated derivatisation procedures will allow high-throughput analysis in proteome research.

References

Annan RS, Köchling HJ, Hill JA, Biemann K (1992) Matrix-assisted laser desorption using a fast-atom bombardment ion source and a magnetic spectrometer. Rapid Commun Mass Spectrom 6:298–302

Barber M, Bordoli RS, Sedgwick RD, Taylor AN (1981) Fast atom bombardment of solids as an ion source in mass spectrometry. Nature 293:270–275

Beavis RC, Chait BT (1989) Matrix-assisted laser-desorption mass spectrometry using 355 nm radiation. Rapid Commun Mass Spectrom 3:436–439

Beavis RC, Chaudhary T, Chait BT (1992) Alpha-cyano-4-hydroxycinnamic acid as a matrix for matrix-assisted laser desorption mass spectrometry. Org Mass Spectrom 27:156–158

Bökelmann V, Spengler B, Kaufmann R (1995) Dynamical parameters of ion ejection and ion formation in matrix-assisted laser desorption/ionization. Eur Mass Spectrom 1:81–93

Brown RS, Lennon JJ (1995) Mass resolution improvement by incorporation of pulsed ion extraction in a matrix-assisted laser desorption/ionization linear time-of-flight mass spectrometer. Anal Chem 67:1998–2003

Castoro JA, Wilkins CL (1993) Ultrahigh resolution matrix-assisted laser desorption/ionization of small proteins by FT mass spectrometry. Anal Chem 65:2621–2627

Chaurand P, Luetzenkirchen F, Spengler B (1999) Peptide and protein identification by Matrix-Assisted Laser Desorption Ionization (MALDI) and MALDI-post-source decay time-of-flight mass spectrometry. J Am Soc Mass Spectrom 10:91–103

Colby SM, King TB, Reilly JP (1994) Improving the resolution of matrix-assisted laser desorption/ionization time-of-flight mass spectrometry by exploiting the correlation between ion position and velocity. Rapid Commun Mass Spectrom 8:865–868

Cordero MM, Cornish TJ, Cotter RJ, Lys IA (1995) Sequencing peptides without scanning the reflectron: post-source decay with a curved-field reflectron time-of-flight mass spectrometer. Rapid Comm Mass Spectrom 9:1356–1361

Doroshenko VM, Cotter RJ (1995) High-performance collision-induced dissociation of peptide ions formed by matrix-assisted laser desorption/ionization in a quadrupole ion trap mass spectrometer. Anal Chem 67:2180–2187

Ehring H, Karas M, Hillenkamp F (1992) Role of photoionization and photochemistry in ionization processes of organic molecules and relevance for matrix-assisted laser desorption ionization mass spectrometry. Org Mass Spectrom 27:472–480

Fenn JB, Mann M, Meng CK, Wong SF, Whitehouse CM (1989) Electrospray ionization for mass spectrometry of large biomolecules. Science 246:64–71

Flad T, Spengler B, Kalbacher H, Brossart P, Baier D, Kaufmann R, Bold P, Metzger S, Blüggel M, Meyer HE, Kurz B, Müller CA (1998) Direct identification of major histocompatibility complex class I-bound tumor-associated peptide antigens of a renal carcinoma cell line by a novel mass spectrometric method. Cancer Res 58:5803–5811

Geno PW, Macfarlane RD (1989) Secondary electron emission induced by impact of low-velocity molecular ions on a microchannel plate. Int J Mass Spectrom Ion Proc 92: 195–210

Jespersen S, Chaurand P, van Strien FJC, Spengler B, van der Greef J (1999) Direct sequencing of neuropeptides in biological tissue by MALDI-PSD mass spectrometry. Anal Chem 71:660–666

Johnson RS, Martin SA, Biemann K (1988) Collision-induced fragmentation of (M + H) + ions of peptides. Side chain specific sequence ions. Int J Mass Spectrom Ion Processes 86:137–154

Karas M, Hillenkamp F (1988) Laser desorption ionization of proteins with molecular masses exceeding 10,000 daltons. Anal Chem 60:2299–2301

Karas M, Bachmann D, Hillenkamp F (1985) Influence of the wavelength in high-irradiance ultraviolet laser desorption mass spectrometry of organic molecules. Anal Chem 57:2935–2939

Karas M, Bachmann D, Bahr U, Hillenkamp F (1987) Matrix-assisted ultraviolet laser desorption of non-volatile compounds. Int J Mass Spectrom Ion Processes 78:53–68

Karas M, Ehring H, Nordhoff E, Stahl B, Strupat K, Hillenkamp F, Grehl M, Krebs B (1991) Matrix-assisted laser desorption/ionization mass spectrometry with additives to 2,5-dihydroxybenzoic acid. Org Mass Spectrom 28:1476–1481

Kaufmann R, Spengler B, Lützenkirchen F (1993) Mass spectrometric sequencing of linear peptides by product ion analysis in a reflectron time-of-flight mass spectrometer using Matrix Assisted Laser Desorption Ionization (MALDI). Rapid Commun Mass Spectrom 7:902–910

Kaufmann R, Chaurand P, Kirsch D, Spengler B (1996) Post-source decay and delayed extraction in MALDI-ReTOF mass spectrometry. Are there trade-offs? Rapid Commun Mass Spectrom 10:1199–1208

Li Y, McIver TJ (1994) Detection limits for matrix-assisted laser desorption of polypeptides with an external ion source Fourier-transform mass spectrometer. Rapid Comm Mass Spectrom 8:743–749

Liao PC, Huang ZH, Allison J (1997) Charge remote fragmentation of peptides following attachment of a fixed positive charge: a matrix-assisted laser desorption/ionization postsource decay study. J Am Soc Mass Spectrom 8:501–509

Nordhoff E, Ingendoh A, Crames R, Overberg A, Stahl B, Karas M, Hillenkamp F, Crain PF (1992) Matrix-assisted laser desorption/ionization mass spectrometry of nucleic acids with wavelengths in ultraviolet and infrared. Rapid Commun Mass Spectrom 6:771–776

Overberg A, Karas M, Bahr U, Kaufmann R, Hillenkamp F (1990) Matrix-assisted infrared-laser (2,94 um) desorption/ionization mass spectrometry of large biomolecules. Rapid Commun Mass Spectrom 4:293–296

Qin J, Chait BT (1996) Matrix-assisted laser desorption ion trap mass spectrometry: efficient isolation and effective fragmentation of peptide ions. Anal Chem 68:2108–2112

Roepstorff P, Fohlman J (1984) Proposal for a common nomenclature for sequence ions mass spectra of peptides. Biomed Mass Spectrom 11:601–601

Spengler B, Cotter RJ (1990) Ultraviolet laser desorption/ionization mass spectrometry of proteins above 100,000 daltons by pulsed ion extraction time-of-flight analysis. Anal Chem 62:793–796

Spengler B, Bahr U, Karas M, Hillenkamp F (1987) Excimer laser desorption mass spectrometry of biomolecules at 248 and 193 nm. J Phys Chem 91:6502–6506

Spengler B, Bahr U, Karas M, Hillenkamp F (1988) Post-ionization of laser-desorbed organic and inorganic compounds in a time-of-flight mass spectrometer. Anal Instrum 17:173–193

Spengler B, Kirsch D, Kaufmann R, Karas M, Hillenkamp F, Giessmann U (1990) The detection of large molecules in matrix-assisted UV-laser desorption. Rapid Commun Mass Spectrom 4:301–305

Spengler B, Kirsch D, Kaufmann R (1991) Metastable decay of peptides and proteins in matrix assisted laser desorption mass spectrometry. Rapid Commun Mass Spectrom 5:198–202

Spengler B, Kirsch D, Kaufmann R (1992) Fundamental aspects of post-source decay in matrix-assisted laser desorption mass spectrometry. J Phys Chem 96:9678–9684

Spengler B, Lützenkirchen F, Kaufmann R (1993) On-target deuteration for peptide sequencing by laser mass spectrometry. Org Mass Spectrom 28:1482–1490

Spengler B, Lützenkirchen F, Metzger S, Chaurand P, Kaufmann R, Jeffery W, Bartlet-Jones M, Pappin DJC (1997) Peptide sequencing of charged derivatives by postsource decay MALDI mass spectrometry. Int J Mass Spectrom Ion Processes 169/170:127–140

Strupat K, Karas M, Hillenkamp F (1991) 2,5-Dihydroxybenzoic acid: a new matrix for laser desorption-ionization mass spectrometry. Int J Mass Spectrom Ion Proc 111:89–102

Van Vaeck L, Gijbels R (1990a) Laser microprobe mass spectrometry: potential and limitations for inorganic and organic micro-analysis, part I. Technique and inorganic applications. Fresenius J Anal Chem 337:743–754

Van Vaeck L, Gijbels R (1990b) Laser microprobe mass spectrometry: potential and limitations for inorganic and organic micro-analysis, part II. Organic applications. Fresenius J Anal Chem 337:755–765

Vestal ML, Juhasz P, Martin SA (1995) Delayed extraction matrix-assisted laser desorption time-of-flight mass spectrometry. Rapid Comm Mass Spectrom 9:1044–1050

Vorm O, Mann M (1994) Improved mass accuracy in matrix assisted laser desorption ionisation time of flight mass spectrometry of peptides. J Am Soc Mass Spec 5:955–958

Whittal RM, Li L (1995) High-resolution matrix-assisted laser desorption/ionization in a linear time-of-flight mass spectrometer. Anal Chem 67:1950–1954

Wiley WC, McLaren IH (1955) Time-of-flight mass spectrometer with improved resolution. Rev of Sci Instrum 26:1150–1157

Wu KJ, Steding A, Becker CH (1993) Matrix-assisted laser desorption time-of-flight mass spectrometry of oligonucleotides using 3-hydroxypicolinic acid as an ultraviolet-sensitive matrix. Rapid Commun Mass Spectrom 7:142–146

4 Data-Controlled Micro-Scale Liquid Chromatography–Tandem Mass Spectrometry of Peptides and Proteins: Strategies for Improved Sensitivity, Efficiency and Effectiveness

Douglas C. Stahl and Terry D. Lee

4.1 Introduction

Over the past decade, electrospray mass spectrometry (ESMS) has emerged as one of the most useful analytical tools for characterising peptides and proteins. As originally conceived, it was a technique that performed best at flow rates of a few microlitres per minute. Such flow rates are significantly lower than those utilised in standard analytical [4.6-mm internal diameter (ID) column] and microbore (2.1-mm ID column) chromatography. As a result, a flurry of activity ensued on the part of mass spectrometer manufacturers in order to design electrospray (ES) interfaces that would accommodate higher flow rates.

The coupling of ESMS to conventional-scale chromatography was certainly useful and was probably essential to its rapid acceptance by the scientific community. However, even at microlitre-per-minute flow rates, the methodology wastes a large portion of the sample. The number of ions detected by the mass spectrometer is generally the same whether the flow rate is 20 nl/min or 20 µl/min, assuming stable spray conditions are achieved in each case. In recent years, many of the technical problems associated with the design of micro-scale chromatographic systems and ES interfaces have been addressed, and liquid chromatography–mass spectrometry (LC-MS) can be readily achieved at flow rates of approximately 100 nl/min. While the pioneering work generally relied on custom-built devices, micro-scale components and systems are now available from a number of commercial vendors. A very powerful and constantly improving technology base is now in place to tackle difficult biological problems involving complex mixtures and very small amounts of sample.

A second and equally important issue is raised by the multi-modal nature of modern tandem mass spectrometry (MS-MS). In addition to precursor scans that provide molecular weight information, it is also possible to obtain structural information on selected ions by collecting MS-MS (fragment-ion) spectra. Optimal use of the sample and the instrument time necessitates that any decision to switch modes of operation be made while the experiment is in progress. Computer control of the data-acquisition process now plays a central role in efficient and effective LC-MS analyses. This chapter describes the basic principles underlying micro-scale ES interfaces, various methodologies for performing on-line micro-capillary chromatography, and the role of computer automation for instrument control and data acquisition.

4.2 The ES Process

ES can be viewed as a two-stage process: the creation of charged droplets from a liquid surface by a strong electric field, followed by evaporation of the charged droplets to yield charged sample molecules. The first stage of the ES phenomenon has been studied for more than 80 years, beginning with the work of Zeleny in 1914 (Zeleny 1914). Most ES emitters consist of hollow, needle-like structures that transport fluid to the concentrated electric field at the tip. In 1964, Taylor proposed an electrostatic model (Taylor 1964) to describe the pointed menisci that form in the liquid from which charged droplets are emitted. Since then, these structures have been referred to as Taylor cones (Fig. 4.1). Investigation of the second stage of the process began in 1968 with the work of Dole and co-workers (Dole et al. 1968). However, it was not until Fenn and colleagues demonstrated the practical application of ES in mass spectrometry (MS) in 1985 that research with biomolecules began (Whitehouse et al. 1985; Fenn et al. 1989). The exact mechanism of ion formation during the ES process is still a matter of some controversy and is beyond the scope of this book. Our discussion of the basic theory will be limited to practical aspects of the methodology.

The stable emission of charged droplets from the tip of a Taylor cone is a function of several parameters, including the distance and potential difference between the emitter and counter electrodes (MS inlet), the emitter dimensions

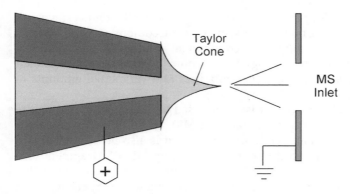

Fig. 4.1. Diagram of an electrospray emitter

and shape, and the sample conductivity and viscosity. Stable ES can be achieved from emitters having relatively large tip dimensions (150 μm), provided flow rates and spray potentials are sufficiently large (2–5 μl/min; 3.5–5 kV). However, changes in solvent composition during liquid chromatography (LC) gradient elution can cause unstable emission, because the required spray voltage is higher for water than for organic solvents. To overcome this limitation, commercial sources operating at these flow rates use a sheath liquid, a nebulising gas or both to achieve stable emission over the entire gradient. This increases the number of parameters that must be optimised empirically, and early versions of commercial ion sources were notoriously finicky. The use of a sheath liquid also dilutes the sample, causing a corresponding decrease in sensitivity.

4.3 Micro-ES Emitter Design

Decreasing the ES-emitter dimensions has many positive benefits. For emitter tips with an outer diameter (OD) of approximately 25 μm, stable emission can be achieved at flow rates as low as 100 nl/min with voltages of approximately 1.5 kV. A sheath liquid and/or nebulising gas is not required. In addition, the emitter can be positioned closer to the inlet of the mass spectrometer so that a greater percentage of sample ions are entrained in the vortex of gas entering the vacuum system. Generally, as the tip size decreases, the emission stability increases. With tip dimensions less than 5 μm, the flow can be reduced to the point that it becomes difficult to measure (below 20 nl/min), and a potential of only 500–1000 V is needed. Smaller ES tips generate smaller initial charged droplets that undergo fewer evaporation-fission cycles before ions are released. As a result, the concentration of components that could suppress sample ionisation is reduced (Juraschek et al. 1998). The main problem with smaller tips is their tendency to occlude. Once occluded, they are difficult to open again. If emitters are to be reused, it is important to filter particles from the solvent stream and to prevent the sample solution from drying in the needles.

4.3.1 Emitter Construction

Although small-diameter metal tubes can be manufactured into ES emitters, the tubing is expensive, difficult to obtain and difficult to connect with other system components. Most micro-scale ES emitters are constructed from fused-silica capillaries (FSCs). FSCs are readily available, relatively inexpensive and come in a variety of sizes. The OD can be as small as 90 μm, and it is possible to spray directly from the end of the tube. However, because solvent tends to wet the entire surface at the end of the tube, the spray characteristics of "blunt-cut" needles are inferior to those with ends that are tapered so that the OD at the tip is only slightly larger than the ID. The OD of the tip can be reduced by simply grinding away the excess material (Kriger et al. 1995; Emmet et al. 1998). Needles constructed in this way from 25-μm ID FSC tubing have ideal properties for many applications. The low internal volume of the needle makes it possible to connect the device to capillary LC columns with little post-column band broadening. The needles are very robust and can be used almost indefinitely. Obstructions are most likely to occur at the entrance of the lumen and can be easily removed by trimming that end. The primary disadvantage of this design is the inability to operate at the very low flow rates required for high sensitivity analyses and specialised chromatographic techniques, such as peak parking (described in later in this chapter).

Smaller tips can be made from FSC tubing by melting the tubing and pulling it to a smaller diameter. This can be done manually using a small gas torch (Davis et al. 1995b) or capillary puller (Davis and Lee 1997). The capillary puller can be programmed to reproducibly generate a variety of different tip shapes. A puller with a laser heat source is required for fused silica tubing and is more expensive than models using heated filaments. With practice and proper technique, the manual method is quite adequate and is very cost effective. FSC tubes with IDs less than 100 μm generally close when pulled. The lumen can be opened by etching the tip with hydrofluoric acid, removing part of the tip by cutting it with a quartz knife or by simply fracturing the end against a hard surface (Gatlin et al. 1998). The etched tips tend to have smooth, symmetrical edges, whereas the other methods tend to yield ragged edges. FSC capillaries with IDs larger than 100 μm can be pulled with a laser puller to form tips with open ends. The shape of the emitter is determined by the laser-power profiles entered into the laser puller. Several commercial vendors now offer FSC emitters in a variety of shapes and sizes.

4.4 High-Voltage Connection

Unlike metal tubes, glass or FSC needles are electrically non-conductive, and a means of connecting an electrical potential to the tip is required. The method employed by Wilm and Mann in their pioneering work on micro-scale ES (nanospray; Wilm and Mann 1994, 1996) was to sputter a metal film onto the outside of the glass needle (Fig. 4.2A). This method avoids the problem of having an electrode in the solvent stream. However, the metal films tend to flake off after prolonged use. Alternative methods have subsequently been described in the literature for creating more permanent metal films on glass surfaces (Kriger et al. 1995; Emmet et al. 1998).

The need to modify the surface of glass or FSC needles can be avoided by using the liquid stream as the conductive path to the emitter. This method, commonly referred to as a "liquid junction", has been used from the very

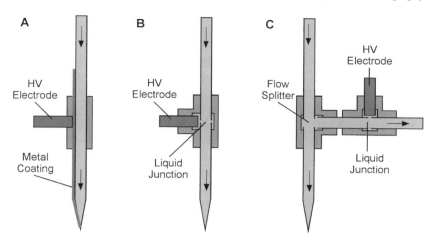

Fig. 4.2A–C. Diagram of different methods for applying an electric potential to micro-scale glass electrospray emitters. **A** Through a metal coating on the outside of the emitter. **B** Via a liquid junction with the solvent stream. **C** Via a liquid junction on the waste side of a split-flow solvent delivery

beginning (Zeleny 1914) and can be achieved in a variety of ways. If an ES emitter is connected to a separation system with a metal union, the union can serve as the electrode in contact with the solvent. If a plastic union or non-conductive ferrules are used, an in-line metal sheath can be used to make contact (Davis et al. 1995a). Alternatively, a tee can be used to place an electrode in contact with the liquid junction (Fig. 4.2B). If the electrical connection is made between the column and the ES needle, it is important to avoid the introduction of any "dead volume" that will degrade chromatographic resolution. With 100-nl/min flow rates, seemingly small volumes become significant. This problem can be avoided by placing the electrical contact upstream from the column, which is mandatory when using columns packed directly in the ES needle.

There are potential complications caused by an electrode in the flow of solvent to the needle. For very conductive solutions, gas bubbles formed at the electrode can disrupt flow and cause intermittent ion emission. Spectral background can increase due to electrochemical reactions with impurities, and peptide modifications, such as the oxidation of methionine residues, can occur. The magnitude of such problems is very sample dependent but, in general, it does not seriously effect the quality of an analysis. The use of gold or platinum electrodes is recommended to avoid the formation of metal-ion adducts with sample molecules. For systems that utilise split flows from the LC pump and off-line column loading, the electrode can be placed in the flow of solvent going to the waste reservoir, thus eliminating bubbles and products from reactions with the electrode (Fig. 4.2C).

4.5 Capillary Columns

A wide variety of capillary columns are available from a number of commercial sources. They generally perform as well or better than larger columns packed with the same support. Since they are expensive, it is advisable to protect them with appropriate filters and to clean dirty samples prior to injecting them. Once occluded with insoluble material, the columns can be difficult to restore. Despite the availability of commercial columns, there are still compelling reasons for making them in the lab. Custom-built columns offer comparable performance at very significant cost savings. There is less hesitancy to discard a homemade column that has marginal performance. For critical applications where cross-contamination must be avoided, they can be considered to be disposable units. The full range of chromatographic supports can be used, and the design can be adapted to a specific analytical need. An excellent example is the very high sensitivity analyses that have been performed using columns packed directly in an ES needle (Emmet and Caprioli 1994).

4.5.1 Column Construction

Columns can be constructed simply and inexpensively from FSC tubing. A membrane filter held in place with a smaller-OD piece of fused silica can be used to retain the chromatographic support (Fig. 4.3A; Davis et al. 1995b). Alternatively, the frit can be positioned between the column and the transfer line within a union (Fig. 4.3B; Ducret et al. 1998). This has been facilitated by the commercial availability of a union assembly with a polymer in-line frit. Another method is to place glass beads in the end of the column and fuse them together to make a frit (Fig. 4.3C; Lemmo and Jorgenson 1993). Such columns can be connected to downstream components with Teflon-tubing connectors if high pressure is not required to obtain flow through the rest of the system. Otherwise, a standard union can be used. The ID of the transfer line should be kept small (25 µm) to avoid post-column band spreading. For most applications, a frit at the beginning of the column is unnecessary. One advantage of FSC columns is the ease with which the top few millimetres can be removed in the event that the top of the column becomes obstructed with particles or precipitated sample. This operation greatly extends the life of a column and eliminates the need for a guard column when analysing samples in complex matrixes.

Capillary columns are generally packed using a suspension of the chromatography media in a suitable solvent. The slurry can be forced into the column using either a pressure bomb (Fig. 4.4A) or a high-performance liquid chromatography (HPLC) pump (Fig. 4.4B). With the pressure bomb, a vial containing the slurry is placed in the bomb, and the inlet of the column is threaded through the lid into the stirred suspension. The system is pressurised with a gas cylinder until the desired amount of packing material has been loaded.

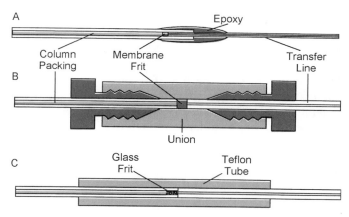

Fig. 4.3A–C. Diagram of different methods for putting a frit on the outlet end of packed capillary columns. **A** A membrane frit on top of a transfer line glued to the end of the column. **B** A membrane frit within a capillary union. **C** Glass beads fused to the end of the capillary column

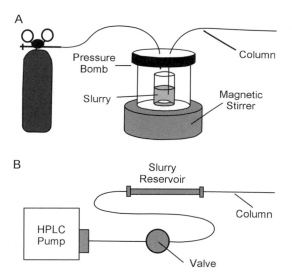

Fig. 4.4A,B. Two different methods of packing capillary columns: (**A**) using a pressure bomb containing a stirred slurry of the chromatography support and (**B**) using a high-pressure liquid pump and a slurry reservoir

With this arrangement, a number of columns can be quickly and easily filled using the same slurry. Alternatively, the slurry can be loaded into a length of stainless-steel tubing connected to an HPLC pump. The pump is pressurised against a closed valve. When the valve is opened, the slurry is "slam packed" into the column. This is the method of choice for making longer columns that require higher packing pressures. The problem of post-column dead volumes can be largely avoided by packing the column directly in the ES needle.

Polymer supports having a uniform bead size (generally 10 μm) can be packed directly into pulled fused-silica needles if the ID of the tip is smaller than the beads (Gatlin et al. 1998). This method does not usually work for silica-based supports, because there are always smaller particles that become tightly wedged in the needle opening and stop the flow. For these materials, it is necessary to place a filter in the tip using a short length of smaller diameter tubing to support the membrane (Davis and Lee 1998). The small void volume between the filter support and the tip is not large enough to be a problem for any but the most demanding applications. Recently, a method for polymerising chromatographic media directly in the ES needle was described (Moore et al. 1998). The resulting monolithic structure has excellent chromatographic properties.

4.6 Gradient-Delivery Systems

Although low flow rate solvent-delivery systems have been introduced by several manufacturers, none are currently capable of delivering sub-microlitre per minute gradients without the use of flow splitting. The simplest split-flow systems use an HPLC tee to divide the stream, sending most of it to waste and the remainder to the capillary column. In theory, the split ratio should remain constant throughout the gradient. In practice, unless the volume ratio of each arm of the tee is the same as the flow ratio, the split ratio will change during the run due to viscosity changes that occur with changing solvent composition. Even larger and less predictable changes occur if the resistance of either arm changes due to flow obstructions created by particles or excess sample that collects on the column. These problems can be alleviated in part by placing a flow resistor between the tee and the column so that only as small portion of the pressure drop is over the column (Chervet et al. 1996; Fig. 4.5A). This general approach is the basis for some commercial split-flow devices.

4.6.1 Flow Splitting

Two different methods have been described for delivering low flow gradients without splitting. The first utilises pre-formed gradients (Ishii et al. 1977). Two low-pressure syringe pumps are used to create a gradient in a loop mounted on a standard six-port valve (Fig. 4.5B; Davis et al. 1995b). The gradient loop is then switched on-line with the flow from a high-pressure syringe pump. To limit the effects of diffusion on the shape of the gradient, it is formed just before it is needed. The system provides excellent separations over a broad range of flow rates and column sizes and allows a variety of variable-flow techniques (see below) that optimise the efficiency of LC-MS analyses.

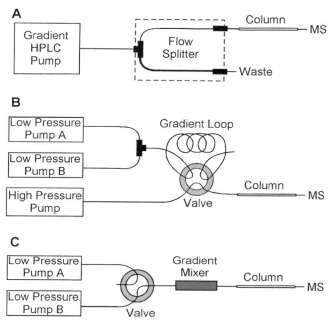

Fig. 4.5A–C. Different methods for obtaining low flow-rate gradients for on-line liquid chromatography–mass spectrometry separations. **A** Split flow from a normal-scale solvent delivery system (the flow splitter is designed so that most of the flow resistance is in the splitter and not the column). **B** Pre-formed gradients formed by two low-pressure pumps and pushed through the column with a high-pressure pump. **C** A mechanical gradient mixer to form exponential gradients in real time

The second non-split solvent delivery system is a version of the exponential-gradient mixer (Takeuchi and Ishii 1982) adapted to nanolitre-per-minute flow rates (Fig. 4.5C; Ducret et al. 1998). The gradient-mixing chamber is a 20-mm length of 0.53 mm OD FSC or Teflon tubing that is initially filled with aqueous solvent. A valve is then switched in order to pump organic solvent into the mixer, which changes the gradient composition in an exponential fashion. A 0.5-mm steel ball is agitated in the chamber to ensure adequate mixing. The low-pressure syringe pumps and mixer design impose limits on the shape and volume of gradients that can be delivered with this system.

4.7 Sample Loading

Standard HPLC injection valves can be used for loading samples onto capillary columns; however, there are a number of potential problems. Larger injection volumes (>1 µl) are generally required, particularly if an auto-sampler is used. At normal capillary-column flow rates, the time required to load several microlitres of sample is significant. The valve creates an additional volume between the pumping system and the column, which substantially delays the start of the gradient. One advantage of the gradient-loop system is that the sample injection valve can be positioned upstream from the gradient valve, which largely eliminates gradient-delivery delays. Fortunately, peptide and protein samples can be loaded at flow rates as high as ten times the normal rate without degrading the separation quality (Davis et al. 1995b). The time saved can be significant, as illustrated by the analysis of a peptide-digest mixture with and without high-speed sample loading (Fig. 4.6). HPLC sample injectors are susceptible to contamination, and the usual precautions are necessary to prevent carry-over from one analysis to the next. There is also a natural tendency to overload capillary columns when working with concentrated samples.

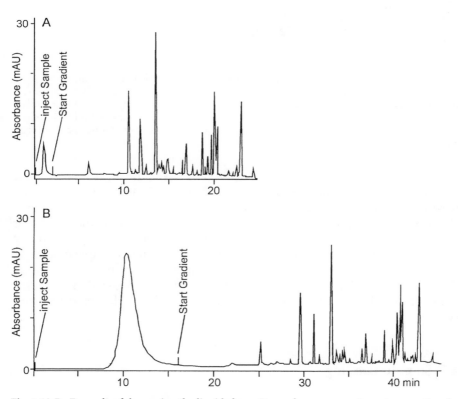

Fig. 4.6A,B. Example of decreasing the liquid chromatography–mass spectrometry run-time by using high-speed sample loading. A mixture of peptides generated by the endo-Lys-c digest of cytochrome C was analysed **A** by loading the sample at ten times the normal elution flow rate and **B** by loading the sample at the normal elution flow rate

The alternative to an injection valve is off-line sample loading with the same pressure-bomb arrangement described for packing columns (Fig. 4.4A; Gatlin et al. 1998). The end of the column can be placed directly into the sample vial, provided the column is long enough to extend through the fitting on the lid of the bomb. The amount of sample loaded is a function of time and pressure and must be empirically determined for each column. Nevertheless, very small sample volumes can be loaded in this manner. The likelihood of contaminating subsequent analyses is greatly reduced and can be totally eliminated by using a new column for each critical analysis. Naturally, this manual method of loading samples would be difficult to adapt to an auto-sampler.

4.8 Variable-Flow Chromatography and Peak Parking

The true power of MS is its ability to provide much more information than merely molecular mass. MS-MS can be used to identify structural moieties from the fragment-ion spectra of selected ion precursors. With sufficient mass accuracy and resolution, elemental compositions can be determined. Components can also be characterised by the loss of specific ionic or neutral fragments. Both qualitative and quantitative data can be obtained. The net result is a tremendous amount of flexibility with which MS can be applied to a particular analytical problem. The process of characterising mixture components as they are presented to a mass spectrometer by a chromatographic separation system must be performed quickly. The most sensitive HPLC analyses separate mixture components into narrow, concentrated bands. As peak width decreases, it becomes less likely that there will be enough time to acquire the desired number and diversity of mass spectra for each eluting component.

One possible solution is to use a very low flow rate, extending the analysis time for all components. This approach is viable for very complex mixtures but would yield a large amount of useless or redundant data for samples where only one or a few regions of the chromatogram are of interest. Another approach is to perform one LC-MS experiment to identify all components that are present, then do as many additional experiments as necessary to collect fragment-ion spectra in selected regions of the chromatogram. This approach requires reproducible chromatography, substantial amounts of sample and prolonged periods of time.

4.8.1 Variable-Flow Chromatography

Changing the traditional relationship between liquid chromatography and MS provides another alternative. When the effluent flow rate is varied in response to feedback from the mass spectrometer regarding the spectral information content, several new analytical options emerge. The flow rate can be increased to minimise the time spent on blank or uninteresting regions of the chromatogram. When a component of interest is detected, the flow can be decreased (peak parking; Davis et al. 1995b) to allow additional time for analysis. In this mode of operation, retention-time information is lost, but the elution order and separation efficiency are retained. This is exemplified by the analysis of a peptide mixture where two components were selected for peak parking (Fig. 4.7). The analysis time for peaks (d) and (e) were extended by more than a factor of five without significantly affecting the peak shapes of components that eluted later. Even peak (g), which was trapped in the flow cell during the peak parking event, eluted as a narrow peak when the normal flow rate was resumed.

The increase in analysis time increases the number and diversity of mass-spectral analysis options. Scan resolution can be increased for specific ions in order to obtain charge-state information. Product-ion spectra can be acquired for different charge states of the same precursor or for ions of multiple, co-eluting precursors. There is also sufficient time for multiple stages of fragmentation on ion-trap and Fourier-transform ion cyclotron-resonance mass spectrometers. Peak parking allows product-ion spectra to be collected while gradually increasing collision energy. A collision-energy gradient can serve two purposes. It provides a mechanism to vary collision energy for the

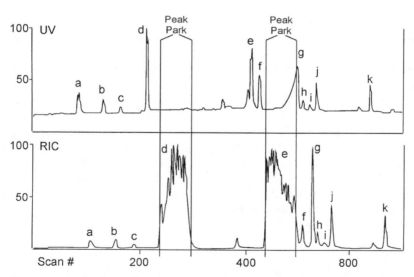

Fig. 4.7. Example of extending the mass-spectrometry analysis times by peak parking. The *upper trace* is the response of an ultraviolet detector connected in series with the electrospray source of a mass spectrometer. The *lower trace* is the base peak chromatogram for ions detected in the mass spectrometer. The two broad peaks in the lower trace are the result of dropping the liquid-chromatography system pressure as these components enter the electrospray interface

generation of optimal spectra at a variety of masses and charge states. In addition, the spectral instability resulting from eventual over-fragmentation of the peptide can be used as a signal to return to precursor-ion scanning (Stahl et al. 1994).

4.8.2 Peak Parking

Variable-flow chromatography facilitates the comprehensive analysis of very complex mixtures with multiple co-eluting components in several regions of the chromatogram (Davis and Lee 1997). A complete set of spectra can be acquired for each component during the course of a single tandem LC-MS experiment. The flow-rate variations and frequent switching between precursor- and product-ion scanning modes give the reconstructed ion chromatogram (Fig. 4.8, upper panel) an unusual appearance but allows an extended analysis of multiple components identified during a single peak-

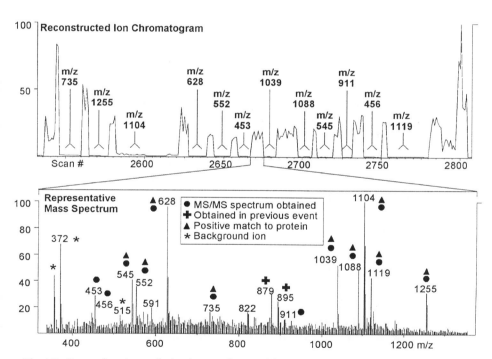

Fig. 4.8. Comprehensive analysis of a complex peptide mixture using variable-flow chromatography. The upper panel is a portion of the reconstructed ion chromatogram corresponding to one peak-parking event. Intensity differences are the result of switching among full-range mass spectrometry (MS) scans, zoom scans and MS-MS scans. The lower panel is a representative full mass-range scan where the components of the extended MS analyses were performed are indicated

parking event (Fig. 4.8, lower panel). When operated in this mode, the HPLC functions like a micro-scale auto-sampler, where components of complex mixtures are presented to the mass spectrometer on demand.

Effective peak parking requires rapid flow-rate changes that cannot be achieved by simply changing the flow rate of a solvent-delivery system operating in a constant-flow mode. Even if the flow rate is rapidly reduced to zero, it takes too long for the system pressure to decrease, and peaks continue to elute until the desired flow rate is achieved. If a valve is added to rapidly decrease column pressure, peak parking can be achieved by switching between two different split ratios. However, it is important that there be no flow restriction between the splitting valve and the column or between the column and the ES needle. Otherwise, it will take too long to achieve the required pressure drop at the needle. The gradient-loop LC system described earlier provides the most flexibility for variable-flow experiments (Davis et al. 1995a). If the high-pressure pump used for gradient delivery can be operated in a pressure-controlled mode, the flow-rate changes required for peak parking can be executed almost instantaneously.

4.8.3 Automation

The comprehensive analysis of complex mixtures may involve dozens of peak-parking events during a single LC-MS experiment. Therefore, automation is required to make the approach practical for routine use. Automated peak parking requires a mechanism to provide the gradient-delivery system with feedback from the mass spectrometer regarding the spectral information content. All mass spectrometers generate electronic signals (analogue or digital) that can be monitored and used for this purpose. These signals are different for each mass spectrometer but can generally be identified on schematics or with assistance from service engineers and the manufacturer's technical support.

Some mass-spectrometer data systems include features that can been used as a platform for the development of interfaces to other instruments (Stahl et al. 1994; Davis et al. 1995b). When this is possible, it can be used to replace the dedicated computer hardware and software required for an HPLC and can facilitate the integration of the systems as a single tandem instrument. When chromatography and MS are under the control of a single data system, an expert system can use information from one instrument to optimise the operation of the other. This level of integration offers several advantages that are described in the following section.

An example of automated peak parking is shown in Fig. 4.9. In response to a precursor ion detected at (1), the system "parked" and collected nine production spectra from a doubly charged peptide. While the HPLC system was still "parked", a single precursor scan detected another peak at (2), and collected ten product-ion spectra for the singly charged species of the same peptide. The

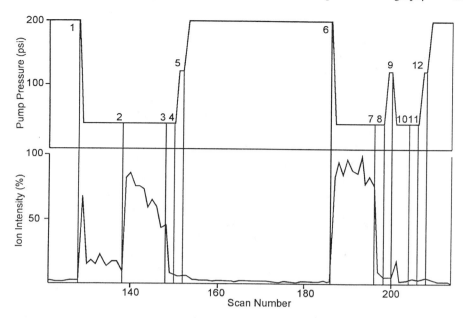

Fig. 4.9. Example of a fully automated variable-flow liquid chromatography–mass spectrometry–mass spectrometry analysis of a peptide mixture. The *upper trace* is a plot of the pump pressure as a function of time (plotted as scan number) during the run. The *lower trace* is the ion intensity recorded by at the mass spectrometer's detector. Points where either the pump pressure or scan mode were changed are *numbered*

system then performed two additional precursor scans to search for additional peptides (3). When none were found, two precursor scans were performed at 50% of the normal HPLC gradient-delivery pressure (4). No new peaks were observed in the interface, so the normal HPLC running pressure was restored (5) until the next peak was observed (6). Events (7) through (12) show the automatic acquisition of product-ion spectra for multiple charge states of a different peptide in the same digest mixture.

4.9 Data-Controlled Analysis

The collection of multiple types of mass spectra on multiple components during the course of a single LC-MS analysis requires the constant modification of instrument parameters in response to data that has just been collected. Such operations are referred to as data-controlled (Stahl et al. 1992, 1996b) or data-dependent (Lim et al. 1999) analyses. Although a human operator can analyse incoming spectra and decide when to change scan modes, the operation is much more efficient and effective when placed under computer control. Most data-controlled LC-MS experiments utilise a "list" of m/z values to make simple decisions about the information found in each precursor-ion spectrum. The base peak of each spectrum is compared with the list to determine if it should be analysed or rejected. If the base peak is rejected, the next most intense peak in the spectrum is evaluated in a similar fashion. This process continues until a user-defined signal-to-noise threshold is reached, at which point a new spectrum is collected for evaluation.

When the analytical goal is to collect product-ion spectra for all unique ion precursors, the experiment generally begins with an empty list. Each unique precursor mass not found in the list is automatically appended to the list before collecting a user-defined maximum number of product spectra for that precursor. Fewer than the maximum number are collected if the signal-to-noise ratio falls below a user-defined minimum. This reduces the risk of missing new precursors while collecting product spectra for a more rapidly eluting component. In this case, the list is used to exclude precursors from further consideration. By forcing the mass spectrometer to search for unique precursors until all have been observed, the depth of analysis is greatly increased. Although mixtures with many co-eluting components may not be fully characterised in a single experiment, the list of precursors observed during a given analysis can be applied to each subsequent analysis until all unique precursors have been identified (Stahl et al. 1996a). In many cases, different precursors with the same m/z values will elute in different regions of the chromatogram. By keeping ions on the exclusion list for a limited time (dynamic exclusion), it is possible to select late-eluting precursors with the same m/z values as those observed earlier in the chromatogram.

4.9.1 Analysing Peptide Modifications

This method is also very useful for locating amino-acid substitutions or post-translational modifications when working with proteins of known sequence. The masses of the expected peptides in a digest mixture can be entered into the exclusion list in advance, forcing the mass spectrometer to collect data only on unknown components. A simple example is the characterisation of a large synthetic protein with an observed mass 64 u higher than expected (Stahl et al. 1996b). The protein was digested with trypsin, and a list of masses for expected tryptic peptides was created. The data-controlled experiment was designed to collect product-ion spectra for *all* precursors that did not have the mass of an expected peptide. Sequence information was obtained for seven peptides in the mixture (Fig. 4.10). The fragment-ion spectrum of one peptide (inset, Fig. 4.10) revealed the substitution of Tyr for Val, accounting for the 64-u difference. The other unknown components were expected peptides observed as metal adducts, disulphide-bonded dimers, and chymotryptic fragments resulting from enzyme contamination.

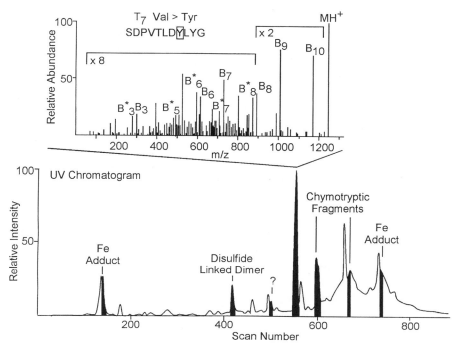

Fig. 4.10. Example of a data-controlled liquid chromatography–mass spectrometry (MS) analysis of the tryptic digest of a small protein. Fragment-ion spectra were collected only for fragments whose observed m/z values did not correspond to those predicted for the expected peptides. The lower trace is the response of an on-line ultraviolet detector. Components for which MS-MS spectra were collected are shaded in *black*. The upper trace is the MS-MS spectra for the peptide having a mass 64-u from the expected value, which established the substitution of Tyr for a Val residue

4.9.2 Software Integration

In general, instruments that are physically coupled to perform tandem separations are not logically linked to interact efficiently and effectively for the solution of challenging analytical problems. Tandem mass spectrometers generally include some degree of hardware and software support for data-controlled analyses. However, the level of support varies greatly among instruments and data systems. Most provide a set of standard procedures with little flexibility for specific user-defined applications.

These limitations can be overcome with a "virtual-instrument" approach to tandem LC-MS analyses. The virtual-instrument paradigm originated with National Instrument's "LabVIEW" (Laboratory Virtual-Instrument Engineering Workbench) software-development platform (Johnson 1994). It uses micro-computers equipped with analogue and digital interface modules to integrate the operation of multiple physical instruments, expand their functionality and allow them to operate as a single, integrated system. LabVIEW can also interact directly with the data systems of commercial instruments via Microsoft Dynamic-Link Libraries (DLLs), Macintosh Code Fragments and UNIX shared libraries. This approach has recently been used to develop an integrated system for the real-time detection and classification of protein bio-markers (Davis et al. 1998; Krishnamurthy et al. 1998).

4.10 Conclusions

Innovations in MS-MS and micro-scale chromatography continue at a rapid pace. A number of labs are developing chip-based systems that will interface with mass spectrometers (Figeys et al. 1997; Ramsey and Ramsey 1997; Xue et al. 1997; Henion et al. 1999; Lebrilla et al. 1999; Licklider et al. 2000). This is expected to further increase the sensitivity and sample throughput. The net result of these and other advances will be a tremendous increase in the quality and quantity of information that can be obtained from complex protein and peptide mixtures. There will also be a corresponding increase in "cognitive overload", which can be minimised through the use of expert systems that automatically apply analytical expertise throughout the course of each analysis.

Expert systems are sophisticated computer programs that manipulate knowledge in order to solve problems (Waterman and Hayes-Roth 1983). In the future, LC-MS expert systems will investigate the complex relationships among incoming data, previously collected data and other sources of information in order to dynamically alter the course of each analysis for the achievement of specific analytical goals. The expert-system approach also provides a means to capture and combine the knowledge of individual experts and transfer it to investigators doing similar work in other laboratories. In short, the approach holds great promise as a necessary bridge between the expertise of researchers who study biological phenomena and the sophisticated analytical tools needed to characterise the molecules responsible for those phenomena. We anticipate that combinations of the strategies and technologies described in this paper will be increasingly applied in many areas of biological research.

Acknowledgements

This work was supported in part by grants from the Public Health Services (NIH RR06217 and CA3572).

References

Chervet JP, Ursem M, Salzmann JB (1996) Instrumental requirements for nanoscale liquid chromatography. Anal Chem 68:1507–1512

Davis MT, Lee TD (1997) Variable flow liquid chromatography tandem mass spectrometry and the comprehensive analysis of complex protein digest mixtures. J Am Soc Mass Spectrom 8:1059–1069

Davis MT, Lee TD (1998) Rapid protein identification using a microscale electrospray LC/MS system on an ion trap mass spectrometer. J Am Soc Mass Spectrom 9:194–201

Davis MT, Stahl DC, Hefta SA, Lee TD (1995a) A microscale electrospray interface for on-line, capillary liquid chromatography tandem mass spectrometry of complex peptide mixtures. Anal Chem 67:4549–4556

Davis MT, Stahl DC, Lee TD (1995b) Low-flow, high-performance liquid chromatography solvent-delivery system designed for tandem capillary liquid chromatography mass spectrometry. J Am Soc Mass Spectrom 6:571–577

Davis MT, Stahl DC, Lee TD (1998) An integrated microspray system for rapid biodetection by mass spectrometry, The 46th ASMS conference on mass spectrometry and allied topics, Orlando, FL, 31 May – 4 June 1998, American Society of Mass Spectrometry, p 284

Dole M, Mach LL, Hines RL, Mobley RC, Ferguson LP, Alice MB (1968) Molecular beams of macroions. J Chem Phys 49:2240–2249

Ducret A, Bartone N, Haynes PA, Blanchard A, Aebersold R (1998) A simplified gradient solvent delivery system for capillary liquid chromatography electrospray ionization mass spectrometry. Anal Biochem 265:129–138

Emmett MR, Caprioli RM (1994) Micro-electrospray mass spectrometry – ultra-high-sensitivity analysis of peptides and proteins. J Am Soc Mass Spectrom 5:605–613

Emmett MR, White FM, Hendrickson CL, Shi S, Marshall AG (1998) Application of micro-electrospray liquid chromatography techniques to FT-ICR MS to enable high-sensitivity biological analyses. J Am Soc Mass Spectrom 9:333–340

Fenn JB, Mann M, Meng CK, Wong SF, Whitehouse CM (1989) Electrospray ionization for mass spectrometry of large biomolecules. Science 246:64–71

Figeys D, Ning YB, Aebersold R (1997) A microfabricated device for rapid protein identification by microelectrospray ion trap mass spectrometry. Anal Chem 69:3153–3160

Gatlin CL, Kleemann GR, Hays LG, Link AJ, Yates JR III (1998) Protein identification at the low femtomole level from silver stained gell using a new fritless electrospray interface for liquid chromatography-microspray and nanospray mass spectrometry. Anal Biochem 263:93–101

Henion J, Heinig K, Wachs T, Schultz G, Corso T (1999) Capillary electrophoresis/mass spectrometry: from one-meter capillaries to chip-based devices. The 47th ASMS Conference on Mass Spectrometry and Allied Topics, Dallas, Texas, 1999. CD-ROM. The American Society of Mass Spectrometry

Ishii D, Asai K, Hibi K, Jonokuchi T, Nagaya M (1977) A study of micro-high-performance liquid chromatography I. Development of technique for miniaturization of high-performance liquid chromatography. J Chromatogr 144:157–168

Johnson GW (1994) LabVIEW graphical programming: practical applications in instrumentation and control. McGraw-Hill, New York

Juraschek R, Dulcks T, Karas M (1998) Nanoelectrospray – more than just a minimized-flow electrospray ionization source. J Am Soc Mass Spec 10:300–308

Kriger MS, Cook KD, Ramsey RS (1995) Durable gold-coated fused silica capillaries for use in electrospray mass spectrometry. Anal Chem 67:385–389

Krishnamurthy T, Davis MT, Stahl DC, Lee TD (1998) Liquid chromatography/microspray mass spectrometry for bacterial investigations. Rap Commun Mass Spectrom 12:1–11

Lebrilla C, Liu J, Tseng K (1999) A micro-fabricated device for coupling capillary electrophoresis and MALDI-MS. The 47th ASMS Conference on Mass Spectrometry and Allied Topics, Dallas, Texas, 1999. CD-ROM, The American Society of Mass Spectrometry

Lemmo AV, Jorgenson JW (1993) Two-dimensional protein separation by microcolumn size-exclusion chromatography-capillary zone electrophoresis. J Chromatogr 633:213–220

Licklider L, Wang XQ, Desai A, Tai YC, Lee TD (2000) A micro-machined chip-based electrospray source for mass spectrometry. Anal Chem 72:367–375

Lim HK, Stellingweif S, Sisenwine S, Chan K (1999) Rapid drug metabolite profiling using fast liquid chromatography, automated multiple-stage mass spectrometry and receptor-binding. J Chromatogr A 831:227–241

Moore RE, Licklider L, Schumann D, Lee TD (1998) A micro-scale electrospray interface incorporating a monolithic, polystyrene-divinylbenzene support for on-line liquid chromatography-tandem mass spectrometry analysis of peptides and proteins. Anal Chem 70:4879–4884

Ramsey RS, Ramsey JM (1997) Generating electrospray from microchip devices using electroosmotic pumping. Anal Chem 69:1174–1178

Stahl DC, Martino PA, Swiderek KM, Davis MT, Lee TD (1992) Automated LC/MS/MS analysis of peptide mixtures using capillary HPLC and electrospray ionization on a triple-sector quadrupole mass spectrometer. The 40th Conference on Mass Spectrometry and Allied Topics, Washington, DC, 1992, American Society of Mass Spectrometry, pp 1801–1802

Stahl DC, Davis MT, Lee TD (1994) Development of a capillary HPLC interface to a Finnigan TSQ 700 mass spectrometer. The 42nd ASMS Conference on Mass Spectrometry and Allied Topics, Chicago, Illinois, 1994, American Society of Mass Spectrometry, p 487

Stahl DC, Davis MT, Lee TD (1996a) Data-controlled modification of the chromatographic time frame using feedback modulation. The 44th ASMS Conference on Mass Spectrometry and Allied Topics, Portland, Oregon, 1996, American Society of Mass Spectrometry, p 1177

Stahl DC, Swiderek KM, Davis MT, Lee TD (1996b) Data-controlled automation of liquid chromatography tandem mass spectrometry analysis of peptide mixtures. J Am Soc Mass Spectrom 7:532–540

Takeuchi T, Ishii D (1982) Continuous gradient elution in micro high-performance liquid chromatography. J Chromatogr 253:41–47

Taylor GI (1964) Disintegration of water drops in an electric field. Proc R Soc Lond A 280:383–397

Waterman DA, Hayes-Roth F (1983) Building expert systems. Addison-Wesley, Reading, MA

Whitehouse CM, Dreyer RN, Yamashita M, Fenn JB (1985) Electrospray interface for liquid chromatographs and mass spectrometers. Anal Chem 57:675–679

Wilm MS, Mann M (1994) Electrospray and Taylor-Cone theory, Dole's beam of macromolecules at last? Int J Mass Spectrom Ion Proc 136:167–180

Wilm M, Mann M (1996) Analytical properties of the nanoelectrospray ion source. Anal Chem 68:1–8

Xue QF, Dunayevskiy YM, Foret F, Karger BL (1997) Integrated multichannel microchip electrospray ionization mass spectrometry: analysis of peptides from on-chip tryptic digestion of melittin. Rapid Commun Mass Spectrom 11:1253–1256

Zeleny J (1914) The electrical discharge from liquid points, and a hydrostatic method of measuring the electric intensity at their surfaces. Phys Rev 3:69–91

5 Solid-Phase Extraction–Capillary Zone Electrophoresis–Mass Spectrometry Analysis of Low-Abundance Proteins

Daniel Figeys and Ruedi Aebersold

5.1 Introduction

Co-ordinated sequencing efforts have already produced the complete genomic DNA sequences of more than 18 prokaryotic and two eukaryotic (*Saccharomyces cerevisiae* and *Caenorhabditis elegans*) species (Goffeau et al. 1996; Blattner et al. 1997; Consortium 1998; see www.tigr.org for more details). However, neither the genomic sequence nor the sequences of expressed genes is sufficient to precisely describe biological processes. This is mainly due to the fact that the level of expression, activity, location in the cell, etc. of the components that constitute such processes are intricately controlled at various levels, including post-transcriptional ones (Gygi and Aebersold 1999). An accurate description of biological processes also requires analysis of the system at the protein level. Proteins are the molecules that control and execute most biological functions, and the level of expression, cellular location and state of activity of a protein are not apparent from its sequence alone.

The most commonly used technique for the separation of complex protein samples is two-dimensional (2D) gel electrophoresis (O'Farrell 1975). The exquisite resolving power of 2D gel electrophoresis has been impressively demonstrated by Klose and co-workers, who were able to separate up to 10,000 components in a single gel (Klose and Kobalz 1995). By itself, gel electrophoresis is a purely descriptive technique. To become useful in proteome analysis, gel electrophoresis had to have the ability to identify the separated proteins. Until recently, the identification of gel-separated proteins essentially relied on chemical sequencing via Edman degradation (Aebersold et al. 1986, 1987). While generally conclusive, this method is not particularly sensitive and is relatively slow. The introduction of more sensitive and potentially faster protein-identification techniques based on mass spectrometry (MS) has significantly enhanced its protein-identification capabilities (Patterson and Aebersold 1995). However, it rapidly became apparent that the quality of the identifications and the sensitivity and throughput achieved critically depend on the quality of the sample that is introduced into the mass spectrometer, irrespective of the type of MS technique used.

Different laboratories therefore attempted to develop improved front-end technologies, such as micro-high-performance liquid chromatography (HPLC; Deterding et al. 1991; Arnott et al. 1993; Davis et al. 1995; Chervet et al. 1996;

Figeys et al. 1998b) and capillary zone electrophoresis (CZE)-based techniques (Figeys et al. 1996a, 1996b; Smith and Udseth 1996; Tomlinson et al. 1996a,c; Bateman et al. 1997; Kelly et al. 1997; Settlage et al. 1998) which, when coupled on-line to electrospray ionisation (ESI) mass spectrometers, significantly enhance the performance of these instruments for protein identification. CZE coupled to MS (Figeys et al. 1996a; Smith et al. 1989, 1991) is a particularly attractive combination, because the two techniques can be directly interfaced, and impressive sensitivities have been achieved in this way.

We were able to successfully analyse mid-attomole to low-femtomole levels of protein standards with a simple CZE–tandem MS (MS/MS) apparatus (Figeys et al. 1996a). However, in CZE, only a small fraction of the sample (nanolitres) is injected onto the column for separation and analysis via MS. This means that a mass limit of detection at the mid-attomole to femtomole level requires an initial sample concentration in the mid-femtomole per microlitre to picomole per microlitre level, which is difficult to achieve for low-abundance proteins. CZE requires a low volume of sample (nanolitres) at relatively high concentration, while enzymatic digestion of proteins produces a dilute mixture of peptides in a large volume (a few microlitres). Furthermore, CZE can be significantly affected by the presence of salts and other matrix components present in samples. Therefore, other workers (Tomlinson et al. 1994, 1996a, 1997) and our group (Figeys and Aebersold 1997, 1998; Figeys et al. 1996a,b, 1997, 1998b) have developed methods to effectively concentrate samples on-line with a CZE separation system so that significantly larger sample volumes can be applied to the capillary separation system.

In our design, termed solid-phase extraction (SPE)-CZE, a small bed of reverse-phase (RP) HPLC material is inserted in front of the CZE and is used for the on-line extraction of peptides. Tens of microlitres can be loaded into the system, eluted from the resin and transferred into the separation capillary with a volume of approximately 100 nl. At the same time, electrokinetic stacking effects were included to further concentrate the eluted sample. Therefore, concentration factors of up to 1000 were obtained, and significant reduction in the chemical background was also achieved due to the ability to extensively rinse the system before elution. SPE-CZE-MS/MS has reached such a degree of maturity that it can be routinely applied for proteome analyses. In this chapter, we provide an overview and a tutorial for the technique.

5.2 Fabrication of the SPE-CZE System

5.2.1 General Design

The SPE cartridges we initially used were built in a manner similar to those already reported (Beattie et al. 1995; Strausbauch et al. 1995, 1996). Over time, we improved and refined their design (Figeys et al. 1996a,b; 1998b). In their

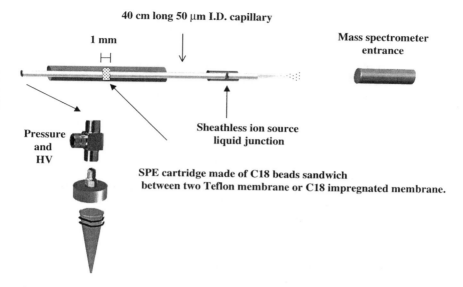

40 cm long 50 μm I.D. capillary

1 mm

Mass spectrometer entrance

Pressure and HV

Sheathless ion source liquid junction

SPE cartridge made of C18 beads sandwich between two Teflon membrane or C18 impregnated membrane.

Fig. 5.1. Diagram of the solid-phase extraction (SPE) capillary zone electrophoresis system coupled to an electrospray-ionisation mass spectrometer. The SPE cartridge is described in more detail in the *insert*

current form (Fig. 5.1; Figeys et al. 1998b), a piece of fused silica capillary [50-μm inner diameter (ID), 150-μm outer diameter (OD), 5 cm in length] is inserted in a Teflon sleeve (250-μm ID or 150-μm ID) and held in place by a hand-tightened fitting. The fitting replaces the procedure previously used (a 5-min application of epoxy). A piece of Teflon membrane is inserted in the other end of the sleeve and is pushed against the capillary. Then, by applying a small vacuum at the capillary end using a syringe fitted with a tapered glass-capillary fitting, a suspension of C18 material (5-μm beads with 300-Å pores) in methanol is pulled into the sleeve. The C18 material is prevented from entering the capillary by the Teflon membrane, and accumulates in the sleeve. After a "column" less than 1 mm in length has formed, the beads are washed with methanol and water. Finally, another piece of Teflon membrane is inserted on top of the C18 material, and the end of the CZE (separation) capillary (50-μm ID, 150-μm OD, 40–60 cm long) is inserted into the sleeve and held in place using a hand-tightened fitting.

This implementation of the SPE method improves some aspects of previous designs, in which the connections between the parts were formed and sealed by glue (Figeys et al. 1996a, 1997). The main advantages of the mechanical connections are the simplicity of assembly, the absence of polymeric contaminants leached out of the glue by organic solvents and the ability to rapidly disassemble and reassemble the cartridge if required. This last factor is important for the removal of air pockets that can be trapped in the void space of the cartridge; such air pockets cause problems with pressurised sample injection. Mechanical connections also allow one to rapidly change defective cartridges without the need to realign the system.

The free end of the CZE capillary is inserted into a clean piece of 180-μm ID stainless-steel tubing (2–3 cm long) and is glued in place using epoxy for 5 min. At the other end of the stainless-steel tube, approximately 3 cm of a 50-μm ID capillary (micro-sprayer) is inserted and glued in place. The micro-sprayer is constructed by using a flame to remove the polyimine coating from approximately 2 cm of the capillary, then pulling the exposed glass in a flame to reduce the outside diameter. The stainless-steel tube, which constitutes a liquid junction for the application of the ESI potential, is mounted on a home-made holder, with the micro-spray tip directed towards the orifice of the mass spectrometer. The mass spectrometer's high-voltage power supply is connected to the stainless-steel tube. High-voltage resistors (a few 5-MΩ resistors) can be added to the cable to limit the maximum amount of current that the mass spectrometer high-voltage power supply can generate. Arcing is unlikely to happen if the stainless-steel liquid junction described in this chapter is used. However, when other methods are used to apply the spraying potential, such as gold-coated capillary ends (Figeys et al. 1996a), the occurrence of arcing is more likely, and it is important to limit the maximal current as a means of protecting the electronics of the mass spectrometer. Using a XYZ translation stage, the tip of the micro-spray end of the capillary is placed approximately 5 mm from the capillary entrance of the mass spectrometer.

We prefer this liquid junction interface because it allows the formation of a robust electrical connection, is easily assembled and allows one to switch between the separation capillary and the micro-sprayer should one of these components become clogged. It has the disadvantages that the dead volume at the junction between the inserted capillaries is variable and unknown, and chemicals can leach out of the epoxy resin used in the liquid junction. This typically causes the appearance of at least one predominant contaminant peak in the MS spectrum. Alternative types of liquid-junction interfaces are possible and have been successfully applied in other applications (Davis et al. 1995; Foret et al. 1996; Wachs et al. 1996; Figeys and Aebersold 1998). For example, we have formed a liquid junction with a low dead-volume micro-connector from Valco that has a theoretical dead volume of only 18 nl. This low dead volume is achieved by a connection that is 1 mm long and has an ID of 150 μm. This means that a linear flow from a 50-μm capillary will be reduced by a factor of nine. Flow reduction at this stage can cause the accumulation of gas generated by the electrolytic process.

5.3 Procedures for the SPE-CZE-MS/MS Experiment

5.3.1 Pre-Saturation

We have found that new RP resins usually contain binding sites that irreversibly absorb peptides. Therefore, for peptide separations at very high sensitivities (low-femtomole to sub-femtomole levels), it is essential to saturate the surface prior to the first use of an SPE column. This is achieved by overloading the SPE cartridge using a mixture of peptides (a tryptic digest of a standard protein) to fully saturate any binding sites. The column is then washed extensively with acetonitrile to remove the reversibly bound analytes and is equilibrated in electrophoresis buffer [acetic acid (10 mM, pH 3.3) and 10% methanol] prior to sample application. If the system is used for higher amounts of sample, this column pre-treatment is not necessary. It is also not necessary to repeat the procedure after the first application, because the resin appears to remain saturated, even after numerous loading/elution cycles.

5.3.2 Sample Loading

Samples are applied to the CZE column by pressure-induced flow through the system. The sample is placed in a pressurisable receptacle, which is then sealed (Fig. 5.1). The pressure applied is manually regulated by a needle valve and is monitored by a manometer connected to the system. Samples are injected by applying 15 psi to the receptacle (Fig. 5.2). Typically, this pressure induces a flow of between 1 and 1.5 µl/min. Care has to be taken to prevent the column from running dry and to avoid the introduction of gas bubbles. Should gas bubbles be inadvertently trapped in the system, they can be easily removed by loosening the fitting and priming the column with electrophoresis buffer if the system is constructed with finger-tightened fittings, as described above. Once the sample is loaded, the apparatus is thoroughly washed with the electrophoresis buffer for 5–10 min at 15 psi. At that point, the pressure is reduced to 0.5 psi, and high voltage (+20 kV) is applied to the system to verify that the system is fully primed and the electrical connection has been maintained. Typically, the current observed with an electrospray process should be approximately 0.8 µA.

5.3.3 Sample Elution and Separation

5.3.3.1 Solid-Phase Extraction–Capillary Zone Electrophoresis

In the simplest form of the technique, peptides are eluted by applying a plug of elution buffer (65% acetonitrile and 3 mM acetic acid) to the column (Figeys

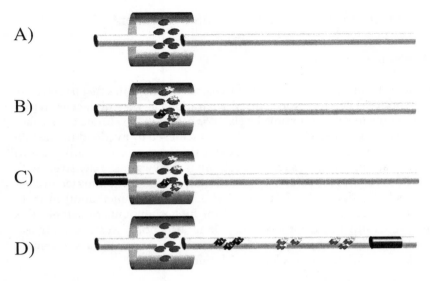

Fig. 5.2A–D. Schematic of the concentration, elution and separation of peptides by solid-phase extraction–capillary zone electrophoresis (SPE-CZE). **A** The SPE-CZE system is rinsed with acetonitrile and electrophoresis buffer [10 mM acetic acid 10% (v/v) methanol] for 5 min at 15 psi. **B** The protein digest is pressure loaded at 10–15 psi, and the system is pressure-rinsed with running buffer for 3–5 min at 10–15 psi. **C** A plug of acetonitrile is pressure injected into the system. **D** Separation is achieved by applying –20 kV and 0.5 psi at the loading end and +1.3 kV at the micro-electrospray interface

et al. 1996a, 1997). The pressurisable sample vessel is filled with elution buffer, and the eluting plug is forced into the capillary by a pulse of positive pressure (6 psi for 6 s). The elution buffer in the sample vessel is then replaced with electrophoresis buffer, and the separation voltage is applied (+20 kV) while a slight positive pressure of 0.5 psi is maintained at the injection end of the capillary. Due to the difference in conductivity between the elution buffer and the electrophoresis buffer, the eluted peptides are further stacked and are then separated. A spraying voltage of +1.8 kV is maintained at the micro-sprayer for the duration of the experiment.

Using this technique, we previously demonstrated the analysis of low-femtomole amounts of peptides isolated after the in-gel digestion of proteins separated by 2D gel electrophoresis (Figeys et al. 1996a, 1997). If the principles of the method are poorly understood, problems can be encountered, especially if the system is constructed from capillaries with uncoated surfaces. The net movement of peptides through the system has two components. The first component is the electrophoretic mobility, which (in this system and with the solvents described) drives the peptides towards the injection end of the capillary. The second component is the electro-osmotic flow, which causes a bulk flow of liquid towards the detector. At a pH of 3.0, the net flow is sufficient to drive the peptides towards the MS detector. However, excessive compression of the SPE cartridge and congestion due to particles present in the sample can restrict the liquid flow. Since the electrophoretic mobility is not affected by f low restriction, the peptides will move back into the SPE cartridge and will be re-adsorbed. In these cases, no analytes will be detected. Increasing the pressure and the injection time can alleviate sample re-adsorption. However, if too much elution buffer is injected, the separation efficiency of the CZE system will decrease. The problem can be eliminated by coating inner surface of the CZE capillary with derivatising agents, such as γ-methacrylopropyl trimethoxysilane or [3-(methacryloylamino)propyl]-trimethylammonium chloride (MAPTAC; Kelly et al. 1997; Figeys et al. 1998b). The effect of this treatment is the introduction of permanent positive charges at the pH used and, therefore, the induction of a strong electro-osmotic flow.

5.3.3.2 SPE and Transient Isotachophoresis

A variation of the basic technique uses transient isotachophoresis during the elution/separation to further concentrate the eluted peptides (Fig. 5.2; Tomlinson et al. 1996b; Figeys et al. 1998a). Elution is performed by injecting a plug of ammonium hydroxide followed by a plug of acetonitrile, using the pressurised sample vessel as described above. The CZE electrophoresis buffer is then installed at the injection end, and the separation voltage is applied (+20 kV and 1.5 psi). The eluted analytes are stacked by transient isotachophoresis and are then electrophoretically separated.

5.3.4 Sample Overloading

Sample overloading is a common problem encountered with the use of very small capillary columns in SPE-CZE-MS/MS systems and in capillary chromatography. Sample overloading typically arises when large amounts of sample or samples contaminated with components that also stick to the column material are applied. In these cases, the small bed of RP material in the SPE cartridge can become saturated. The analytes with the highest affinity for the RP material will displace analytes of lower affinity; the result is that a few specific analytes are accumulated and concentrated on the column, while other compounds are not bound and are washed out. Figure 5.3 illustrates this displacement-chromatography effect.

The SPE cartridge was overloaded with a peptide mixture at high concentration. Peptides were then eluted and analysed by CZE-MS/MS. We observed that the initial peaks, which often represent the most hydrophobic analytes (lower-charge-state ions in solution), have high intensity, while the peptides detected later (i.e. the peptides with stronger electrophoretic mobilities) are detected at significantly decreased intensities. Sample overloading is easily

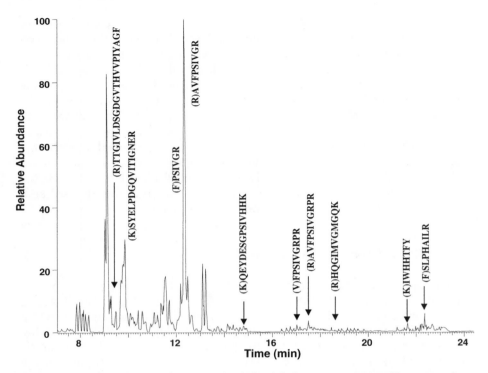

Fig. 5.3. Separation obtained for an overloaded solid-phase extraction–capillary zone electrophoresis–mass spectrometry system. Analysis of 670 fmol at 45 fmol/μl of ACT yeast obtained from a two-dimensional polyacrylamide gel in which a total yeast cell lysate had been separated. The peaks that contain peptides from ACT yeast are labelled with the peptide sequence

overcome if the sample applied is diluted to reduce the sample load and the high level of contaminants. This typically brings the amount of analyte to within the capacity range of the SPE cartridge and increases the number of peptides that are retained on the SPE cartridge.

5.3.5 MS and Data Processing

The rapid identification of minute amounts of proteins on a SPE-CZE system (or any separation system coupled on-line to an electrospray mass spectrometer) requires the rapid and automated generation of collision-induced dissociation (CID) spectra (also referred to as fragment-ion spectra or MS/MS) as soon as the analytes elute from the separation system. This type of experiment is called data dependent MS/MS or "on-the-fly" MS/MS. It requires the detection of the peptide ions in an MS scan followed by the isolation and fragmentation of selected peptides by CID to generate a MS/MS spectrum from which partial or complete peptide sequence information can be extracted. Obviously, such experiments require a tandem mass spectrometer than can operate in a data-dependant manner. Specifically, the instrument needs to decide which of the ions detected in the MS scan is selected for CID, determine the optimal CID conditions for the selected peptide ion, switch from MS to MS/MS mode, select the specific peptide ion for CID, perform the CID experiment and record the fragment-ion spectra. This process is then repeated with a different ion. To avoid repeated analysis of the same ion (a contaminant present at a high concentration), the masses of the ions already analysed are added to a list, which prevents them from being selected again during the same experiment. Therefore, the instrument requires a high degree of programming flexibility and the capacity to rapidly and automatically process data during acquisition. Currently, only a handful of mass spectrometers have this ability. We have predominantly used the TSQ 7000 triple quadrupole and the LCQ ion-trap mass spectrometers (Ducret et al. 1996; Figeys and Aebersold 1997) from Finnigan Mat to perform SPE-CZE-MS experiments. Other mass spectrometers, such as the ion-trap MS from Bruker, the Qtof-MS from Micromass and the Qstar-MS from Sciex, either offer or will soon offer a similar programming capability. SPE-CZE is a generic method that can be easily fitted to any type of ESI-MS instrument. However, to fully utilise the power of the technique, mass spectrometers with the ability to perform data-dependent MS/MS experiments should be utilised.

Protein identification based on CID spectra has been dramatically accelerated by the availability of sequence databases and the development of computer algorithms that correlate the information contained in CID spectra with sequence databases. Protein databases are increasingly available and can be generated using a three-frame translation of a DNA-sequence database or a six-frame translation of expressed-sequence tag databases. The different software systems either identify the provenance of peptides using uninterpreted MS/MS spectra (SEQUEST; Eng et al. 1994; Yates et al. 1995, 1996) or partially interpreted MS/MS spectra [using programs such as Sequence tag (Mann and Wilm 1994), Prowl (Fenyo et al. 1998) and Protein Prospector (Baker and Clauser 1996)]. If several MS/MS spectra generated from peptides derived from the same protein correspond to different sections of that protein, unambiguous sequence identity is usually established.

We have developed a decision scheme for the rapid processing of multiple CID spectra generated in SPE-CZE-MS/MS experiments. The scheme is illustrated in Fig. 5.4. The underlying idea is to progress from the automated batch processing of spectra to more labour-intensive and time-consuming analyses if automated data analysis provides no or inconclusive results. First, the SEQUEST software (Eng et al. 1994; Yates et al. 1996) is used to search databases automatically and rapidly with limited user intervention. If the results are inconclusive, sequence tags can be manually generated and used with various database-searching programs (Mann and Wilm 1994; see http://www.matrixscience.com for links to the mascot program). At that

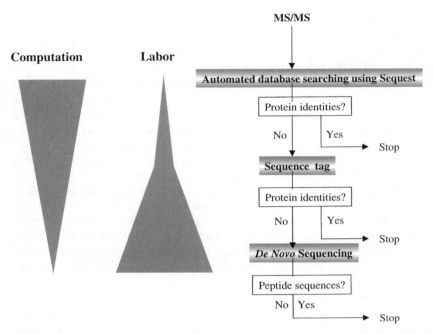

Fig. 5.4. Protein and DNA-sequence database-searching scheme for data generated using the solid-phase extraction–capillary zone electrophoresis–tandem mass-spectrometry protocol

point, it is worthwhile to consider the question of whether it is more time effective to manually interpret the spectra or to attempt to generate better-quality data, which might lead to the automatic identification of the protein. Finally, de novo sequencing is used if no matches are established by database searching (if the spectra are of sufficient quality for de novo sequencing). De novo sequencing is typically performed manually. However, for good-quality data, the Lutefisk software (Taylor and Johnson 1997) available at http://www.lsbc.com:70/Lutefisk97.html can facilitate the process.

5.4 System Optimisation and Variations of the Basic Method

5.4.1 Extraction Materials for the SPE Cartridge

Different RP materials can be utilised for the SPE cartridge. However, the user should be aware that they do not all perform equally. RP beads are generally designed for optimal performance in HPLC experiments. Even though similar solvents are used, the requirements for optimal performance in RP-HPLC and SPE-CZE experiments are different, mostly due to the fact that, in CZE, the beads are exposed to an electric field. Generally, we found the RP materials to perform well during the sample-loading phase of the SPE-CZE experiment. However, we observed differences in the extraction of basic peptides and, with some types of beads, substantial problems arose during peptide elution and separation. The quality of the RP resin used for the construction of the SPE cartridge is of critical importance for the success of the SPE-CZE experiment. The criteria a resin must meet in order to perform well in an SPE-CZE are different from those it must meet for chromatography applications. There are a number of reasons for this. First, the stationary phase has to remain stable and must remain on the resin in an electric field. We found that some types of beads shed the stationary phase in an electric field. Second, analytes have to be efficiently extracted from a relatively large sample volume, typically corresponding to many column volumes, and must be retained during extensive washing with electrophoresis buffer. Third, due to the small size of the column, resins need to have a high binding capacity to avoid saturation with contaminants and displacement of analytes (overloading). Fourth, the resin needs to be compatible with sample elution in very small volumes to maintain high analyte concentrations for CZE. Small elution volumes can compromise the elution, because the organic solvent injected is diluted with sample buffer present in the pores of the beads and because the RP material will absorb some of the organic solvent (as would any hydrophobic analyte). Finally, the resin should have good flow characteristics, thus minimising the impedance of the electro-osmotic flow, especially when bare fused-silica capillary tubes are used.

We tested four different RP materials for suitability in SPE cartridges. The resins were Spherisorb (5-μm beads, 300-Å pore size), VYDAC (5-μm beads, 300-Å pore size), YMC gel ODS-AQ (5- to 15-μm beads, 120-Å pore size) and Monitor (5-μm beads, 300-Å pore size; Figeys et al. 1998b). We prepared SPE cartridges with the four RP resins, and the cartridges were successively connected to the same MAPTAC-coated separation capillary and the SPE-CZE-MS system. Sample elution was performed under standard conditions (acetonitrile elution, no transient isotachophoresis). One of the RP materials (VYDAC) rapidly failed when exposed to the elution solvent and electric field. We detected high levels of released polymers via MS. Therefore, this material was excluded from further analysis.

For the remaining materials, three criteria were considered. The first was the ability of the RP materials to extract peptides from the sample and retain them during the injection and rinsing processes. We first compared the RP materials with regard to the amount of peptide accumulated on the beads if a constant amount of peptide was injected. Three different peptides [fibronectin-related peptide (FRP), α-endorphin 1 and tyrosine-bradykinin (Y-brad)], which differ in their pIs, were used in these experiments, and the amount of peptide absorbed (based on the signal obtained from the eluted peptides) was measured. Figure 5.5A shows a histogram plot of the amount of peptide retained by each one of the reversed phase materials used, normalised to the

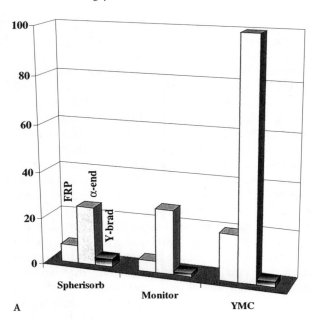

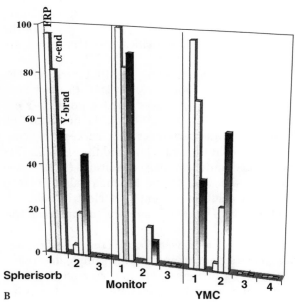

Fig. 5.5A,B. Effect of different extraction materials on **A** the accumulation of peptides on the solid-phase extraction cartridge and **B** the sequential elution of peptides. α-*end* α-Endorphin 1, *FRP* fibronectin-related peptide, *Y-brad* tyrosine-bradykinin

highest value. It is apparent that both the Spherisorb and Monitor material performed similarly, while the YMC material performed better for medium- to high-pI peptides.

The second criterion we applied was the fraction of peptide released during a typical elution process. Figure 5.5B shows a normalised histogram plot of the signals obtained by serial elution of peptides from the SPE-CZE-MS system. The Monitor material released most of the peptide in the first elution cycle. The Spherisorb material exhibited similar elution properties, although a larger fraction of the peptide eluted during the second elution cycle. The YMC mate- rial exhibited strongest peptide retention. Four elution cycles were required to completely elute the peptides.

For the third criterion, we evaluated peak broadening caused by the RP resins. We observed that the peak widths obtained for the Spherisorb and the Monitor materials were similar, while the peak widths were significantly increased for the YMC material. We attributed this observation to the strong peptide-retention properties of this material. Altogether, these results indicate that the highest accumulation of peptides was obtained for the YMC material, followed by the Spherisorb and the Monitor materials. The first two criteria indicate that the YMC packing is more suited for our type of application and that, due to the strong peptide retention, at least two elution CZE-MS/MS cycles can be performed from one injection of sample. The same experiments were repeated for the YMC and Spherisorb resins under transient- isotachophoresis conditions. The SPE cartridges were connected to bare fused- silica capillaries. We quickly noticed that the YMC beads leaked polymers under these conditions, while the Spherisorb material remained stable and performed well.

5.4.2 Surface Chemistry for Capillary Tubing

When samples containing small amounts of peptide are analysed, sample loss due to adsorption to wetted surfaces is a concern. Depending on the nature of the surface and the protein/peptide, analytes can be completely absorbed. The charge states of the inner surface of the separation capillary affect both reso- lution and sample recovery. The number and density of the surface charges primarily determine the direction and strength of the electro-osmotic flow and, thus, the resolution. Sample recovery is reduced, sometimes dramatically, if an analyte and the capillary inner surface carry opposite charges; the analyte is thus electrostatically adsorbed to the capillary. Furthermore, capillary tubes have a high surface-to-volume ratio. Therefore, the probability of interaction between dissolved analytes and surfaces is great. Bare fused-silica capillaries have free silanol groups present on their inner surface. The pK$_a$ values of these groups are approximately 3.8–5. The different values can be explained by the fact that the silanols are immobilised and close enough to interact, which means that the protonation or de-protonation of one group can affect the pK_a values of neighbouring groups. Therefore, the measured pK_a value is dependent on the density of free silanol groups on the surface.

Typically, SPE-CZE experiments are performed in electrophoresis buffer at a pH of between 3.0 and 3.4. This range is slightly below the pK_a values of the silanol groups. Therefore, most of the silanol groups on the capillary wall are present as SiOH and, depending on the pK_a value used, 1–39% of the group are in the form SiO^-. Interactions between positively charged peptides and the negatively charged inner wall of the capillary causes significant band broadening and reduces the recovery of some peptides due to adsorption to the capillary. Furthermore, as described above, the slight negative surface charge of uncoated capillaries generates an electro-osmotic flow towards the anode (ion source), whereas the analytes electrophoretically migrate towards the cathode (injection end). Under normal operating conditions, the electro-osmotic mobility is slightly higher than the electrophoretic mobility of even the fastest-migrating analytes; thus, peptides are detected in the MS. If the electro-osmotic flow is impeded, however, the electrophoretic mobility of at least some analytes exceeds the electro-osmotic flow. Consequently, these analytes migrate back towards the injection end of the capillary and are eventually re-absorbed by the SPE device. Impedance of the electro-osmotic flow could occur via compression of the C18 resin in the SPE cartridge, misalignment of the separation capillary and the micro-electrospray interface in the liquid junction (Fig. 5.1), or via flow constriction in the micro-electrospray needle.

To reduce problems related to the surface charges of capillaries, we have utilised two different capillary coatings that both provide an amino group attached to the inner wall of the capillary tube through a spacer. The immobilised amino groups are protonated at a pH of 3.0 and provide a positive surface that repels peptides with pIs greater than 3.0. The procedures for obtaining the coatings are described in the protocol section at the end of this chapter. The 3-amino-propylsilane-coated capillary (Figeys et al. 1996a) provides high electro-osmotic pumping, and most analytes elute within 5 min from a 40-cm-long capillary at 475 V/cm (Fig. 5.6A). The MAPTAC-coated capillary (Figeys et al. 1998b) provides lower electro-osmotic pumping and better separation of analytes (Fig. 5.6B) and, hence, exhibits better performance.

5.4.3 Separation

The power of the SPE-CZE system is the combination of efficient sample concentration and high-resolution separation. The RP cartridge in which peptides are absorbed from a large sample volume achieves sample concentration. During the SPE-CZE-MS/MS experiment, the peptides accumulated on the RP material are released using a small amount of acetonitrile. Since the mass spectrometer is essentially a concentration-dependent device, the sensitivity will increase with increasing sample concentration. Transient isotachophoresis, a method in which the eluted peptides are further concentrated by electrophoretic stacking, has been reported to significantly reduce the peak width

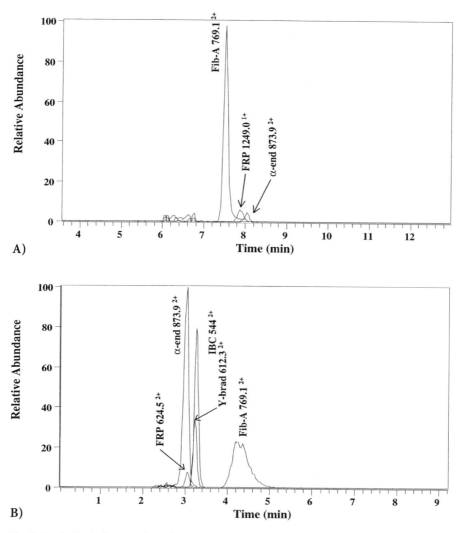

Fig. 5.6A–C. Typical separation of a mixture of five peptides on solid-phase extraction (SPE) capillary zone electrophoresis (CZE) tandem mass-spectrometry systems with **A** uncoated capillaries, **B** 3-aminopropyl silane-coated capillaries and **C** [3-(methacryloylamino)propyl]-trimethylammonium chloride-coated capillaries. A Spherisorb C18-packed SPE column and a 40-cm-long CZE capillary (50-μm inner diameter and 150-μm outer diameter) were used in all cases. The peptide mixture contained α-endorphin 1 (24 fmol/μl), fibrinopeptide A (32 fmol/μl), fibronectin-related peptide (28 fmol/μl), insulin β-chain 22–30 (36 fmol/μl) and tyrosine-bradykinin (35 fmol/μl). The experiments were performed using **A** 9 μl, **B** 6 μl and **C** 9 μl of the peptide mixture. Elution was performed by injecting acetonitrile (65%) and 3-mM acetic acid for 20 s at 4 psi. The separation was performed by applying −19 kV and 1 psi to 10 mM acetic acid and 10% methanol at the injection end and +1.3 kV at the micro-electrosprayer

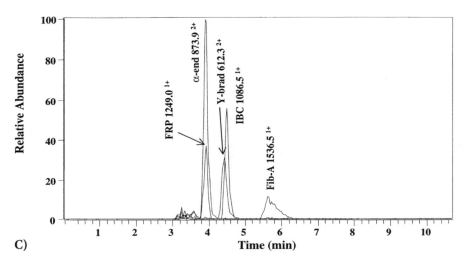

C)

Fig. 5.6A–C. *Continued*

in CZE experiments (Tomlinson et al. 1996b, 1997) and is, therefore, expected to increase the sensitivity of SPE-CZE-MS/MS experiments. During isotachophoresis, a leading electrolyte with high mobility is added before the elution buffer, and the running buffer is used as the lower-mobility, trailing electrolyte. In order to perform this technique on the SPE-CZE-MS/MS system, the leading electrolyte has to be compatible with analysis by MS. Ammonium hydroxide was reported as a suitable leading electrolyte (Smith et al. 1996; Tomlinson et al. 1996b, 1997).

We have tested the effects of transient isotachophoresis in an SPE-CZE-MS/MS system consisting of an uncoated capillary and a SPE cartridge built containing YMC or Spherisorb C18 material. In these experiments, 10 μl of a peptide mixture was pressure loaded on-line onto the SPE cartridge. The SPE cartridge was washed with electrophoresis buffer. This was followed by the sequential injection of a leading electrolyte (0.1% ammonium hydroxide for 10 s at 4 psi) and of elution buffer (65% acetonitrile and 3-mM acetic acid for 20 s at 4 psi). The electrophoresis buffer was then placed at the injection end of the capillary, and the separation was started by applying −19 kV and 1 psi. We quickly noticed that the YMC resin leaked polymers under these conditions, so YMC was no longer used for these experiments. If the SPE cartridge was constructed from Spherisorb resin, we noticed that the temporal peak width was not significantly improved by transient isotachophoresis (Fig. 5.7A). However, the spatial peak width was improved significantly, because electro-

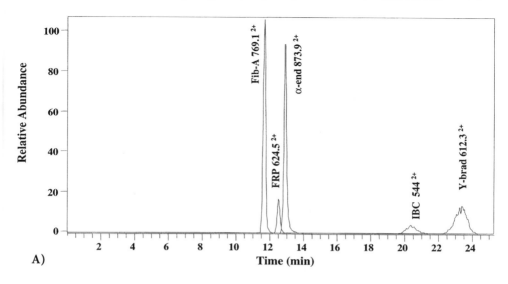

A)

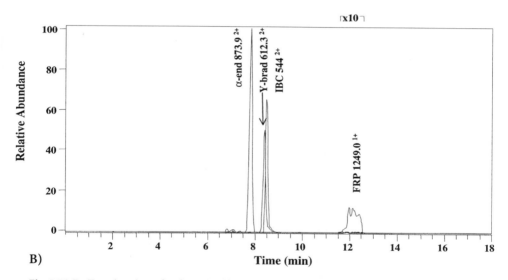

B)

Fig. 5.7A,B. Transient isotachophoresis effect on the separation of the peptide mixture using **A** uncoated and **B** [3-(methacryloylamino)propyl]-trimethylammonium chloride (MAPTAC)-coated solid-phase extraction (SPE) capillary zone electrophoresis (CZE) tandem mass spectrometry. The SPE cartridge was made with Spherisorb C18 material. The uncoated CZE capillary was 40 cm long, and the MAPTAC-coated CZE capillary was 80 cm long. The experiments were performed using **A** 9 μl and **B** 6 μl of the peptide mixture described in the caption of Fig. 5.5. Elution was performed by injecting ammonium hydroxide (0.1%, pH 11) for 10 s at 4 psi and acetonitrile (65%) and 3-mM acetic acid for 20 s at 4 psi

osmotic pumping was reduced by the procedure. The observed peak height increased by up to twenty-fold, and all the peptides in the peptide mixture were observed. The basic peptides Y-brad and insulin β-chain 22–30 were observed as very broad peaks. Sometimes, we noticed the presence of ammonium–peptide adducts.

We also performed experiments with transient isotachophoretic sample concentrations on SPE-CZE-MS/MS systems with MAPTAC-coated capillaries. We discovered that the volume of ammonium hydroxide had to be kept minimal to keep the coating from peeling off the inner wall of the capillary. Apparently, ammonium hydroxide at a pH of 11 destroys the coating. This was apparent from the detection of polymers in experiments in which higher loads of ammonium hydroxide were applied. Similar to capillaries with an MAPTAC coating, 3-amino-propylsilane-coated capillaries were also damaged if an electric field was applied with a solvent having a pH higher than five. However, both coatings were stable when the injected volume of ammonium hydroxide was kept low (10 s, 4 psi). Figure 5.7B shows the separation of the peptide mixture described above on an MAPTAC-coated capillary using ammonium hydroxide as the leading electrolyte. On this system, fibrinopeptide A was not detected, and FRP was detected as a broad, low-intensity peak. Both peptides have low pIs, suggesting that the unfavourable chromatographic properties were caused by strong electrostatic interactions with the wall of the coated capillaries.

5.4.4 Peak-Parking Mode

The duty cycles of some mass spectrometers (including triple-quadrupole and ion trap mass spectrometers) limit the number of tandem mass spectra that can be acquired during the elution of a specific peak from the SPE-CZE-MS system. Typically the MS/MS instruments are programmed to isolate peptide ions for CID in order of decreasing intensity. One analytical cycle requires approximately 1–3 s and consists of peptide-mass analysis, peptide-ion selection and acquisition of the MS/MS spectra derived from the selected peptide ion(s). Generally, only the one to three most intense peptide ions are analysed during a typical SPE-CZE-MS experiment, leaving the less intense peptide ions unanalysed. This is not a problem for the identification of a single protein by database searching, because complete coverage of the peptide map is not required. However, this can limit the analysis of complex samples generated by the digestion of protein mixtures and the analysis of the post-translational modification of proteins present at low stoichiometry. Particularly in the latter case, it is essential that the peptides of lowest abundance be included in the analysis.

This problem could be alleviated if the elution time of a peak could be extended. The concept of the "reduced elution speed" of peptides in CZE-MS describes an approach to increase the number of peptides for analysis by MS (Goodlett et al. 1993). Referred to as "peak parking", this method extends the analysis time available for specific analytes by reducing the sample flow from the peptide separation system to the mass spectrometer. Peak parking has been successfully implemented for capillary HPLC (Davis et al. 1995; Davis and Lee 1997), CZE (Smith et al. 1996; Ducret et al. 1996) and, recently, SPE-CZE-MS analysis (Figeys et al. 1999; Gallis et al. 1999).

5.4.4.1 Instrument-Control Procedures

Routine performance of the peak-parking procedure requires the automation of the instrument-control system and data-dependent control of the peptide-separation parameters. Currently, this can only be achieved on a few types of mass spectrometers that have the capability of advanced programming. We have developed data-dependent peak-parking procedures for the TSQ 7000 and the LCQ ion-trap MS from Finnigan Mat. On the TSQ 7000, we utilised the instrument-control language (ICL) capabilities to write a set of five programs that collectively control the TSQ-7000 instrument, the acquisition of CID spectra and the SPE-CZE high-voltage power supply. To implement data dependent peak parking on the LCQ mass spectrometer, we utilised the extended functions of a beta software version from Finnigan MAT, a home-built controller that taps into the LCQ hardware and a protocol written in LabVIEW that controls the peak-parking procedure. Both systems perform essentially the same functions and are described below.

For the first 5 min (or longer, if no ions are detected) of an SPE-CZE-MS/MS experiment, the MS was scanned in MS mode only, and the CZE separation was performed at high voltage (−20 kV). After the initial 5 min, any ion that was detected in the MS with an intensity exceeding a pre-set threshold value (and for which no CID spectrum had been acquired in the same experiment) triggered the peak-parking mode. Once committed to the peak-parking mode, the system automatically decreased the flow of analytes from the CZE by increasing the voltage applied to the upstream end (resulting in a decrease in the potential difference over the capillary) of the capillary (cathode) from −20 to −5 kV, while the voltage applied to the micro-sprayer was constant. The mass spectrometer was also instructed to switch from the MS mode to the MS/MS mode and to generate and record CID spectra. If no ions were detected or if all the ions detected in a peak were subjected to CID, the instrument continued to scan in MS mode, and the electrophoretic mobility was not reduced. Refer to Figeys et al. (1999) and Gallis et al. (1999) for more details on the whole process. The ICL procedures are freely available from http://weber.u.washington.edu/~ruedilab/aebersold.html or http://www.garvan.unsw.edu.au/public/corthals. They can be directly pasted into the ICL window of any TSQ-7000 instrument.

5.4.4.2 Analysis of Protein with the Peak-Parking Mode

We have tested the SPE-CZE-MS system under peak-parking conditions with a set of standardised protein digests. In these experiments, we achieved limits of detection similar to those previously reported for SPE-CZE-MS/MS (Figeys et al. 1996a, 1997). The utility of the peak-parking procedure was also demonstrated for the analysis of the sites of in vivo phosphorylation of the enzyme endothelial nitric-oxide synthase (eNOS), which was isolated from bovine aortic endothelial cells by immunoprecipitation. The determination of the eNOS in vivo phosphorylation sites was complicated by the low abundance of the enzyme, the low stoichiometry of phosphorylation and the large size of the protein, which resulted in a large number of peptide fragments after enzymatic digestion. The isolated phosphoprotein was separated from contaminating proteins by gel electrophoresis and was digested with trypsin, and the phosphopeptides were enriched by immobilised-metal affinity chromatography, as previously described (Watts et al. 1994; Corthals et al. 2000). The fraction containing the phosphopeptides was further separated on a micro-bore RP-HPLC column.

The analysis of one of the fractions by SPE-CZE-MS/MS, using the peak-parking procedure and the isotachophoretic concentration (Sect. 5.3), is illustrated in Fig. 5.8A. The resulting CID spectra were searched against the OWL protein database using the SEQUEST software (Eng et al. 1994; Yates et al. 1995). In this fraction, two peptides were matched to eNOS, one of which was phosphorylated. Another five peptides in the same fraction were matched to keratin. The other CID spectra generated were either not of high enough quality to permit unambiguous peptide identification or were derived from small, singly charged ions that did not contain sufficient information to enable them to be assigned by the SEQUEST software. The same fractions processed by other separation techniques and by SPE-CZE without peak parking failed to identify the phosphorylated peptides. Figure 5.8B shows the change in voltage applied to the cathodic end of the SPE-CZE system during the peak-parking experiment.

5.5 Conclusion

SPE-CZE-MS has become one of the most sensitive techniques for the analysis of low-abundance proteins and can easily be performed with a limited budget. We have demonstrated various applications of the SPE-CZE technique for protein identification and characterisation. Further incremental improvements will increase the performance and robustness of the method, and the use of resins with specific selectivities for the construction of the SPE cartridge will allow the specific enrichment of selected analytes from complex sample mixtures. Automation of the SPE-CZE-MS/MS process will be required in order

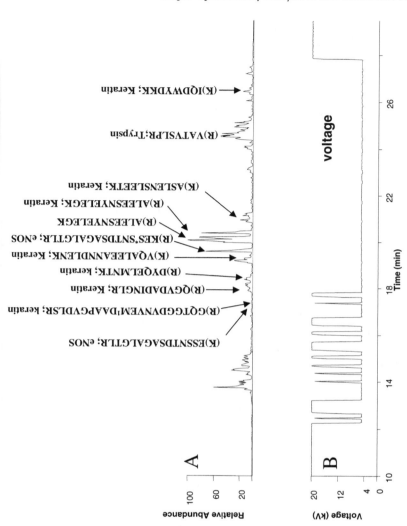

Fig. 5.8A,B. Analysis by solid-phase extraction–capillary zone electrophoresis–tandem mass spectrometry (SPE-CZE-MS/MS) in peak-parking mode. Peptides derived by tryptic digestion of endothelial nitric oxide synthase were separated by reverse-phase high-performance liquid chromatography (HPLC). Peptides contained in a single HPLC fraction were further separated by SPE-CZE-MS/MS under peak-parking conditions. **A** The base peak trace for the analysis. **B** The voltage modulations (flow modulation) of the SPE-CZE power supply during the experiment. The sequence of the peptides and their provenance are identified on the spectra

to make the technique useful as a general analytical tool in the field of proteomics. We believe that the technique has matured to such an extent that it can be implemented with limited training. We hope to transfer the technique to other laboratories so that it will be more widely applied.

Appendix: Protocols

1 SPE-CZE (Quick Guide)

1. Rinse the system with 100% acetonitrile at 15 psi for 5 min.
2. Rinse the system with electrophoresis buffer (10 mM acetic acid, 10% methanol) at 15 psi for 5 min.
3. Load the sample at 15 psi. Do not allow the system to dry out.
4. Rinse the system with electrophoresis buffer at 15 psi for 5 min.
5. Elute the sample with 65% acetonitrile and 3 mM acetic acid at 6 psi for 6 s.
6. Run the experiment with electrophoresis buffer at +20 kV and 0.5 psi.

2 SPE-CZE-Transient Isotachophoresis (Quick Guide)

1. Rinse the system with 100% acetonitrile at 15 psi for 5 min.
2. Rinse the system with electrophoresis buffer (10 mM acetic acid, 10% methanol) at 15 psi for 5 min.
3. Load the sample at 15 psi. Do not allow the system to dry out.
4. Rinse the system with electrophoresis buffer at 15 psi for 5 min.
5. Inject 0.1% ammonium hydroxide at 4 psi for 10 s.
6. Elute the sample with 65% acetonitrile and 3 mM acetic acid at 4 psi for 10–20 s.
7. Run the experiment with running buffer at +20 kV and 0.5 psi.

3 Coating of Capillaries: General Comments

It is preferable to prepare a few metres of capillary tubing at a time. For a 50-µm capillary, 5 m can be easily prepared. However, care must be taken to extensively rinse the capillary to avoid plugging. Omission of rinsing between the sodium hydroxide hydrochloric acid steps will cause salt precipitation at the interface between the two solutions and plugging of the capillary. The two first coatings are easy to perform and provide Si–O–Si attachments to the capillary tubing. However, their stability, once wet, is limited to a few days due to the hydrolysis of the Si–O–Si attachments to the capillary wall. The coatings are stable if the capillaries are stored dry. Basic solutions readily remove both the γ-methacrylopropyl trimethoxysilane and the MAPTAC coatings with Si–O–Si anchors. Alternative coating procedures based on Grignard reagents are presented and provide a Si–C–Si attachment to the wall for use under conditions that require enhanced stability under alkaline conditions. Care must be exercised in handling reagents employed in the coating procedures, as they are generally corrosive and may react with water (vinyl magnesium bromide and thionyl chloride).

4 γ-Methacrylopropyl Trimethoxysilane Coating

1. Rinse the capillary with sodium hydroxide 0.1 M for 30 min at 20 psi.
2. Rinse with water for 10 min at 20 psi.
3. Rinse with hydrochloric acid (1 M) for 30 min at 20 psi.
4. Dry the capillary at 110 °C overnight with 5–10 psi of helium. This is necessary to remove any traces of water, which would react with the reagent in step 6.
5. Rinse with toluene for 10 min at 20 psi.
6. Pass a 10% (w/v) solution of 3-aminopropyltrimethoxysilane in toluene through the capillary at 30 psi for 3 h (this reagent reacts easily with water).
7. Bake the capillary at 103 °C overnight, maintaining a flow of helium through the capillary at 20 psi. This step helps to produce a more stable coating. The final coated capillary can be kept dry for years. However, once aqueous solutions are utilised, the coating generally degrades within a few days due to hydrolysis. Basic solutions will rapidly remove the coating. The coating will systematically fail when the capillaries are subjected to a solution with a pH above 5.0 or an electric field (Bruin et al. 1989; Thorsteinsdottir et al. 1995; Figeys et al. 1996a).

5 MAPTAC Coating with Si–O–Si Attachments

1. Rinse the capillary with sodium hydroxide (0.1 M) for 30 min at 20 psi.
2. Rinse with water for 10 min at 20 psi.
3. Rinse with hydrochloric acid (1 M) for 30 min at 20 psi.
4. Pass a solution of 7-oct-1-enyltrimethoxysilane (50 μl) and glacial acetic acid (50 μl) in 10 ml of methanol for 3 h at 20 psi. The 7-oct-1-enyltrimethoxysilane is available from Gelest (Tullytown, Penn., USA). The 7-oct-1-enyltrimethoxysilane anchors itself to the inner wall of the capillary tubing and provides a vinyl group for further derivatisation.
5. Rinse with methanol, followed by water at 20 psi for 10 min.
6. Prepare a fresh solution of 2% (v/v) MAPTAC in 10 ml water. Add 20 μl of N,N,N′,NN′-tetramethylethylenediamine (TEMED) and 140 μl of ammonium persulphate (10% w/v).
7. Pass the solution through the capillary at 20 psi for 3 h.
8. Purge the column with water, dry with nitrogen and store (Kelly et al. 1997).

6 MAPTAC Coating with Si–C Attachments

1. Rinse the capillary with sodium hydroxide (1.0 M) for 30 min at 20 psi.
2. Rinse with water for 30 min at 20 psi.
3. Rinse with methanol for 30 min at 20 psi.
4. Flush with nitrogen overnight at 120 °C.
5. Flush the capillary with thionyl chloride from a chamber pressurised at 20 psi and 70 °C until the flow is steady. Thionyl chloride is acidic; use litmus paper to verify flow. *Caution: thionyl chloride is a lachrymator. Do not allow the column effluent to empty into the lab.*
6. Reduce the pressure to 2 psi and flush with thionyl chloride for 12 h at 70 °C.
7. Replace the thionyl chloride with 1-M vinyl magnesium bromide in tetrahydrofuran (THF).
8. Apply 20 psi until the vinyl magnesium bromide appears at the capillary exit, then reduce the pressure to 2 psi for 6 h.
9. Replace the Grignard reagent with THF and apply 20 psi until a few milli-litres have passed through the capillary.
10. Repeat step 9 with water.
11. Flush the capillary with nitrogen.
12. Prepare a fresh solution of 2% (v/v) of MAPTAC in 10 ml. Add 20 µl of TEMED and 140 µl of ammonium persulphate (10% w/v) and pass the solution through the capillary at 20 psi for 30 min at 70 °C.
13. Purge the column with water, dry with nitrogen and store (Dolnik et al. 1998).

References

Aebersold R, Teplow D, Hood L, Kent S (1986) Electroblotting onto activated glass. High efficiency preparation of proteins from analytical sodium dodecyl sulfate-polyacrylamide gels for direct sequence analysis. J Biol Chem 261:4229–4238

Aebersold R, Leavitt J, Saavedra R, Hood L, Kent S (1987) Internal amino acid sequence analysis of proteins separated by one- or two-dimensional gel electrophoresis after in situ protease digestion on nitrocellulose. Proc Natl Acad Sci USA 84:6970–6974

Arnott D, Shabanowitz J, Hunt D (1993) Mass spectrometry of proteins and peptides: sensitive and accurate mass measurement and sequence analysis. Clin Chem 39:2005–2010

Baker PR, Clauser KR (1996) Protein prospector. http://prospector.ucsf.edu, San Francisco

Bateman KP, White RL, Thibault P (1997) Disposable emitters for on-line capillary zone electrophoresis nanoelectrospray mass spectrometry. Rapid Commun Mass Spectrom 11:307–315

Beattie JH, Self R, Richards MP (1995) The use of solid phase concentrators for on-line precon-centration of metallothionein prior to isoform separation by capillary electrophoresis. Electrophoresis 16:322–328

Blattner FR, Plunkett GP III, Bloch CA, Perna NT, Burland V, Riley M, Collado-Vides J, Glasner JD, Rode CK, Mayhew GF, et al. (1997) The complete genome sequence of Escherichia coli K-12. Science 277:1453–1474

Bruin G, Huisden R, Kraak J, Poppe H (1989) Performance of carbohydrate-modified fused-silica capillaries for the separation of proteins by zone electrophoresis. J Chromatogr 480:339–349

Chervet JP, Ursem M, Salzmann JB (1996) Instrumental requirements for nanoscale liquid chromatography. Anal Chem 68:1507–1512

Consortium TCeS (1998) Genome sequence of the Nematode C. elegans: a platform for investigating biology. Science 11:2012–2018

Corthals GL, Gygi SP, Aebersold R, Patterson SD (2000) Identification of proteins by mass spectrometry. In: Rabilloud T (ed) Proteome research: 2D gel electrophoresis and detection methods. Springer, Berlin Heidelberg New York, pp 197–231

Davis MT, Lee TD (1997) Variable flow liquid chromatography-tandem mass spectrometry and the comprehensive analysis of complex protein digest mixtures. J Am Soc Mass Spectrom 8:1059–1069

Davis MT, Stahl DC, Hefta SA, Lee TD (1995) A microscale electrospray interface for on-line, capillary liquid chromatography/tandem mass spectrometry of complex peptide mixtures. Anal Chem 67:4549–4556

Deterding LJ, Parker CE, Perkins JR, Moseley MA, Jorgenson JW, Tomer KB (1991) capillary liquid chromatography-mass spectrometry, and capillary zone electrophoresis-mass spectrometry for the determination of peptides and proteins. J Chromatogr 554:329–338

Dolnik V, Xu D, Yadav A, Bashkin J, Marsh M, Tu O, Mansfield E, Vainer M, Madabhushi R, Barker D et al (1998) Wall coating for DNA sequencing and fragment analysis by capillary electrophoresis. J Microcolumn 10:175–184

Ducret A, Bruun CF, Bures EJ, Marhaug G, Husby G, Aebersold R (1996) Characterization of human serum amyloid A protein isoforms separated by two-dimensional electrophoresis by liquid chromatography/electrospray ionization tandem mass spectrometry. Electrophoresis 17:866–876

Eng J, McCormack AL, Yates JR III (1994) An approach to correlate tandem mass spectral data of peptides with amino acid sequences in a protein database. J Am Soc Mass Spectrom 5:976–989

Fenyo D, Qin J, Chait BT (1998) Protein identification using mass spectrometric information. Electrophoresis 19:998–1005

Figeys D, Aebersold R (1997) High sensitivity identification of proteins by electrospray ionization tandem mass spectrometry: initial comparison between an ion trap mass spectrometer and a triple quadrupole mass spectrometer. Electrophoresis 18:360–368

Figeys D, Aebersold R (1998) High sensitivity analysis of proteins and peptides by capillary-electrophoresis-tandem mass spectrometry: recent developments in technology and applications. Electrophoresis 19:885–892

Figeys D, Ducret A, van Oostveen I, Aebersold R (1996a) Protein identification by capillary zone electrophoresis-microelectrospray ionization-tandem mass spectrometry at the subfemtomol level. Anal Chem 68:1822–1828

Figeys D, Ducret A, Yates JRIII, Aebersold R (1996b) Protein identification by solid phase microextraction-capillary zone electrophoresis-microelectrospray-tandem mass spectrometry. Nat Biotechnol 14:1579–1583

Figeys D, Ducret A, Aebersold R (1997) Identification of proteins by capillary electrophoresis tandem mass spectrometry – evaluation of an on-line solid-phase extraction device. J Chromatogr A 763:295–306

Figeys D, Gygi SP, Zhang Y, Watts J, Gu M, Aebersold R (1998a) Electrophoresis combined with novel mass spectrometry techniques: powerful tools for the analysis of proteins and proteomes. Electrophoresis 19:1811–1818

Figeys D, Zhang Y, Aebersold R (1998b) Optimization of solid phase microextraction-capillary zone electrophoresis-mass spectrometry for high sensitivity protein identification. Electrophoresis 19:2338–2347

Figeys D, Corthals GL, Gallis B, Goodlett DR, Ducret A, Corson MA, Aebersold R (1999) Data-dependent modulation of solid-phase extraction capillary electrophoresis for the analysis of complex peptide and phosphopeptide mixtures by tandem mass spectrometry: application to endothelial nitric oxide synthase. Anal Chem 71:2279–2287

Foret F, Kirby DP, Vouros P, Karger BL (1996) Electrospray interface for capillary electrophoresis mass spectrometry with fiber-optic UV detection close to the electrospray tip. Electrophoresis 17:1829–1832

Gallis B, Corthals GL, Goodlett DR, Ueba H, Kim F, Presnell SR, Figeys D, Harrison DG, Berk BC, Aebersold R, Corson MA (1999) Identification of flow-dependent endothelial nitric oxide synthase phosphorylation sites by mass spectrometry of phosphorylation and nitric oxide production by the PI3-kinase inhibitor LY294002. J Biol Chem 274:30101–30108

Goffeau A, Barrell BG, Bussey H, Davis RW, Dujon B, Feldmann H, Galibert F, Hoheisel JD, Jacq C, Johnston M et al (1996) Life with 6000 genes. Science 274:546, 563–567

Goodlett DR, Wahl JH, Udseth HR, Smith RD (1993) Reduced elution speed detection for capillary electrophoresis-mass spectrometry. J Microcolumn 5:57–62

Gygi S, Aebersold R (1999) Absolute quantitation of 2-D protein spots. Methods Mol Biol 112:417–421

Kelly JF, Ramaley L, Thibault P (1997) Capillary zone electrophoresis-electrospray mass spectrometry at submicroliter flow rates: practical considerations and analytical performance. Anal Chem 69:51–60

Klose J, Kobalz U (1995) Two-dimensional electrophoresis of proteins: an updated protocol and implications for a functional analysis of the genome. Electrophoresis 16:1034–1059

Mann M, Wilm M (1994) Error-tolerant identification of peptides in sequence databases by peptide sequence tags. Anal Chem 66:4390–4399

O'Farrell PH (1975) Hgih resolution 2D gel electrophoresis of proteins. J Biol Chem 250:4007–4021

Patterson SD, Aebersold R (1995) Mass spectrometric approaches for the identification of gel-separated proteins. Electrophoresis 16:1791–1814

Settlage RE, Russo PS, Shabanowitz J, Hunt DF (1998) A novel μ-ESI source for coupling capillary electrophoresis and mass spectrometry: sequence determination of tumor peptides at the attomole level. J Microcolumn 10:281–285

Smith RD, Udseth HR (1996) Capillary electrophoresis/mass spectrometry. In: Lunte SM, Radzik DM (eds) Pharmaceutical and biomedical applications of capillary electrophoresis, vol 2. Pergamon, New York, pp 229–275

Smith RD, Loo JA, Barinaga CJ, Edmonds CG, Udseth HR (1989) Capillary zone electrophoresis and isotachophoresis-mass spectrometry of polypeptides and proteins based upon an electrospray ionization interface. J Chromatogr 480:211–232

Smith RD, Udseth HR, Barinaga CJ, Edmonds CG (1991) Instrumentation for high-performance capillary electrophoresis-mass spectrometry. J Chromatogr 559:197–208

Smith RD, Udseth HR, Wahl JH, Goodlett DR, Hofstadler SA (1996) Capillary electrophoresis-mass spectrometry. In: Karger BL, Hancock WS (eds) Methods in enzymology, vol 271. Academic Press, New York, pp 448–486

Strausbauch MA, Madden BJ, Wettstein PJ, Landers JP (1995) Sensitivity enhancement and second-dimensional information from solid phase extraction-capillary electrophoresis of entire high-performance liquid chromatography fractions. Electrophoresis 16:541–548

Strausbauch MA, Landers JP, Wettstein PJ (1996) Mechanism of peptide separations by solid phase extraction capillary electrophoresis at low pH. Anal Chem 68:306–314

Taylor JA, Johnson RS (1997) Sequence database searches via de novo peptide sequencing by tandem mass spectrometry. Rapid Comm Mass Spec 11:1067–1075

Thorsteinsdottir M, Isaksson R, Westerlund D (1995) Performance of amino-silylated fused-silica capillaries for the separation of enkephalin-related peptides by capillary zone electrophoresis and micellar electrokinetic chromatography. Electrophoresis 16:557–563

Tomlinson AJ, Benson LM, Braddock WD, Oda RP (1994) On-line preconcentration-capillary electrophoresis-mass spectrometry (PC-CE-MS). J High Resol Chromatogr 17:729–731

Tomlinson AJ, Benson LM, Guzman NA, Naylor S (1996a) Preconcentration and microreaction technology on-line with capillary electrophoresis. J Chromatogr A 744:3–15

Tomlinson AJ, Benson LM, Jameson S, Naylor S (1996b) Rapid loading of large sample volumes, analyte cleanup, and modified moving boundary transient isotachophoresis conditions for membrane preconcentration-capillary electrophoresis in small diameter capillaries. Electrophoresis 17:1801–1807

Tomlinson AJ, Jameson S, Naylor S (1996c) Strategy for isolating and sequencing biologically derived MHC class I peptides. J Chromatogr A 744:273–278

Tomlinson AJ, Benson LM, Jameson S, Johnson DH, Naylor S (1997) Utility of membrane preconcentration capillary electrophoresis mass spectrometry in overcoming limited sample loading for analysis of biologically derived drug metabolites, peptides, and proteins. J Am Soc Mass Spectrom 8:15–24

Wachs T, Sheppard RL, Henion J (1996) Design and applications of a self-aligning liquid junction electrospray interface for capillary electrophoresis mass spectrometry. J Chromatogr B 685:335–342

Watts JD, Affolter M, Krebs DL, Wange RL, Samelson LE, Aebersold R (1994) Identification by electrospray ionization mass spectrometry of the sites of tyrosine phosphorylation induced in activated Jurkat T cells on the protein tyrosine kinase ZAP-70. J Biol Chem 269:29520–29529

Yates JR III, Eng JK, McCormack AL, Schieltz D (1995) Method to correlate tandem mass-spectra of modified peptides to amino-acid-sequences in the protein database. Anal Chem 67:1426–1436

Yates JR III, Eng JK, Clauser KR, Burlingame AL (1996) Search of sequence databases with uninterpreted high-energy collision-induced dissociation spectra of peptides. J Am Soc Mass Spectrom 7:1089–1098

6 Protein Identification by Peptide-Mass Fingerprinting

PAOLA DAINESE and PETER JAMES

6.1 Introduction

The idea that a set of (poly)peptide masses obtained by specific enzymatic or chemical cleavage can be used as a unique fingerprint, allowing a protein to be identified in a database, was first proposed in 1977 (Cleveland et al. 1977). The masses of the polypeptides generated by in-gel proteolysis (estimated from sodium dodecyl sulphate polyacrylamide-gel electrophoresis gels) were used to identify viral-coat proteins. The mass accuracy obtained by gel electrophoresis was sufficient to allow identification, because the size of the database was so small. Later, the combination of peptide mapping and mass spectrometry (MS) was introduced as a method for verifying and/or correcting the primary structures of proteins, as deduced from their corresponding gene sequence (Gibson and Biemann 1984; Morris et al. 1984; Greer et al. 1988). The strategy implied the use of fast atom bombardment (FAB) or plasma-desorption MS to determine the masses of the proteolytic fragments produced by a specific enzyme or chemical. The experimental data was then compared with the masses predicted from all three reading frames of the gene sequences in order to verify the protein sequence or show the presence of post-translational modifications. Greer demonstrated that FAB mapping allowed the confirmation of 93% of the sequence of recombinant 1-anti-trypsin molecules (Greer et al. 1988); this introduced the use of MS-based quality control of expressed proteins. They also showed that the N-terminus of the protein was acetylated at methionine, since an unexpected tryptic fragment of $m/z = 1231$ disappeared after cyanobromide (CNBr) treatment, giving rise to a new signal of $m/z = 1058$ (consistent with the loss of acetyl-Met). The method seemed powerful and required what for that time was a very small amount of material (nanomoles).

6.1.1 Protein Identification Using Proteolytic Masses

In a poster presentation at the third Symposium of the Protein Society in Seattle in 1989, Bill Henzel re-introduced the idea of peptide-mass finger-printing (PMF) as a rapid alternative to peptide sequencing for protein identification. The concept appeared to remain unused until 1993, when five

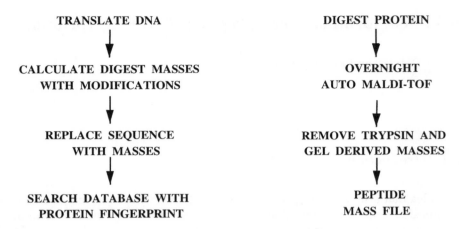

Fig. 6.1. An outline of the process of peptide-mass fingerprinting: generating real and virtual data for comparison

independent laboratories published methods and algorithms for database searching using mass-spectral data (Henzel et al. 1993; James et al. 1993; Mann et al. 1993; Pappin et al. 1993; Yates et al. 1993). The same general approach was common to all the papers. The general strategy of the method is outlined in Fig. 6.1. First, a protein is digested by a chemical or a protease with high sequence specificity (such as CNBr or trypsin) to produce a set of peptides. The molecular masses are determined by matrix-assisted laser desorption ionisation (MALDI)-MS or electrospray-ionisation MS and form the mass fingerprint. This experimentally determined set of masses is subsequently used to search a set of mass profiles generated by theoretical fragmentation of a protein database in order to find the protein generating the most similar pattern. The method is now popularly called PMF.

The computer search produces a list of proteins matching – at least in part – the experimental data. The mass tolerance allowing a match between a calculated mass value and an experimental one is set either by the operator or by the algorithm itself. The procedure is fast and, under given conditions, is very reliable. All five groups stressed the importance of PMF as a mean of linking two-dimensional (2D) gel-spot databases to protein databases. Currently, several software tools that provide PMF searching of sequence databases using various types of mass-spectral data are available through the World Wide Web or by e-mail (Table 6.1).

Table 6.1. e-mail addresses and World-Wide Web servers providing peptide-fingerprint database-searching tools

Program	e-mail address/website
MOWSE	e-mail server, use the word "HELP" in the message body and send to mowse@dl.ac.uk
Protein prospector	http://prospector.ucsf.edu/
MASCOT	http://www.matrixscience.com
DARWIN	http://cbrg.inf.ethz.ch/MassSearch.html
PeptideSearch	http://mac-mann6.embl-heidelberg.de/
Prot-ID	http://chait-sgi.rockefeller.edu/cgi-bin/prot-id
Expasy tools	http://www.expasy.ch/tools/

6.2 Database Searching and Scoring Schemes

The original five papers described different algorithms for database searching, each with different, user-selectable parameters for calculating the list of the best matching proteins. Patterson (Patterson and Aebersold 1995) performed a comparison and evaluation of the various database-searching algorithms. Three of the algorithms (Henzel et al. 1993; Mann et al. 1993; Yates et al. 1993) use a similar scoring scheme to order the proteins according to the decreasing number of matching peptides. The approximate molecular weight (MW) range of the intact protein can be used as a pre-filter to eliminate random matches due to the presence of very large proteins in the database. However, this method assumes that the protein being analysed does not differ greatly in mass from the predicted mass in the database. This eliminates the ability to identify highly glycosylated proteins or proteolytic fragments. Yates et al. circumvented this problem by using the MW as a sliding window and only recording masses that occur within that range. Thus, if the analysed sample is a fragment of a larger protein, the program counts only the matches in a consecutive region of the larger protein corresponding to the estimated MW introduced by the operator.

In MOWSE (Pappin et al. 1993) and MassSearch (James et al. 1993), the scoring schemes are based on probability. In the first case, a matrix of weighting factors is calculated during the database's creation; proteins are initially grouped into 10-kDa MW intervals and, inside each interval, the MWs from the digested proteins are assigned to cells of 100-Da intervals. For each 10-kDa interval, the cell-frequency values are then normalised with respect to the maximum frequency value determined, to give final values between zero and one. During the database search, each matching peptide yields a frequency value from the calculated frequency plot and, in cases of multiple matches for a protein, the frequency values are multiplied. For the computa-

Fig. 6.2. Peptide-mass fingerprinting and probability: balls-in-a-box analogy

tion of the final score, this number is then inverted and normalised to an average protein mass of 50 in order to prevent large random scores caused by large proteins.

A different approach was used in the MassSearch algorithm, which provides the user with an objective measure of the statistical significance of a match. Fig. 6.2 shows a container of unit length, inside of which are k small boxes of length ε, which are the peptide masses from a theoretical digest of a protein in the database. Several balls representing the MWs of peptides obtained from the digestion of the unknown are placed in the container. One can calculate the chance that the balls will randomly land in each box.

Suppose k small intervals (or boxes) are selected, each with a length $n\varepsilon$ in the range zero to one, and ε is small enough so that the k intervals distribute at random without a significant danger of overlap. A length $k\varepsilon$ of the interval is then covered, and $1 - k\varepsilon$ is uncovered. For all practical purposes, the term $(1 - \varepsilon^{-n\varepsilon})^k$ is an excellent approximation of the probability. The masses that give the lowest probability are selected from the set. This will happen for k^* of the weights (given k weights in the sample, where n is the number of weights for the entire digested protein). The logarithms of the weights are taken and normalised to a unit interval by division by $\log(w_{max}) - \log(w_{min})$, where w_{max} and w_{min} are the highest and lowest measured weights, respectively. For each experimental mass, the closest calculated mass in the database is identified and, for a given number of selected weights k^*, a radius is defined (representing the largest miss). For this k^* and this radius, a probability $Pr(\varepsilon,k^*,n)$ can be computed; the final score is calculated as $10 \times \log_{10}[Pr(\varepsilon,k^*,n)]$, the probability that the matches occurred at random. The calculated probability P is then converted to $(-\log_{10}P)$ and is the score in the list of matching proteins. This enables a score to be defined; below this score, the noise in the database is too high, and the match should be considered insignificant.

6.3 Strengths and Limitations of the Method

In addition to showing the advantages of this new method, the original papers emphasised some of potential limitations of the approach. The importance of the mass accuracy achieved was highlighted; the tighter the mass tolerance, the more stringent the identification. Fewer peptides were needed to get a high score (James et al. 1993), and one did not need to specify an upper limit for the protein MW (Mann et al. 1993). The effect of mass accuracy (0.01 vs. 0.2%)

Table 6.2. Mass accuracy and database searching. All spectra were obtained using pulsed extraction in reflectron mode without internal standards

Score	n	k	Accession number	Description/species
Mass accuracy = ±1.5 (Δ = 15.1)				
96.1	13	8	P06117	Nitrogenase iron protein, *Bradyrhizobium japonicum*
81.0	7	3	P31363	Octamer-binding trans factor 11, *Mus musculus*
80.9	29	4	P36953	Afamin precursor, *Rattus norvegicus*
74.4	17	3	P25740	Lipopolysaccharide synthase, *Escherichia coli*
Mass accuracy = ±0.2 (Δ = 96.1)				
172.6	13	8	P06117	Nitrogenase iron protein, *Bradyrhizobium japonicum*
118.1	9	5	P00463	Nitrogenase iron protein, *Bradyrhizobium japonicum*
80.1	12	3	P46416	Glutathione synthase, *Arabidopsis thaliana*
76.4	56	4	P25391	Laminin α-1, *Homo sapiens*

on the PMF is demonstrated in Table 6.2 using the MassSearch program (James et al. 1993). The correct matching protein is identified even when the instrument is tuned to low resolution.

However, due to the greater inaccuracy, some proteins – especially larger ones – appear in the list due to random matches, and the score values of the correct proteins drastically drop. Most important, the difference (Δ) between the score of the correct protein and the next, non-related protein is small. This value (rather than the absolute score) is the most important factor in determining the confidence that can be attached to the search result. The initial reports stressed the wide mass-accuracy tolerance of the method because, at that stage, the protein databases being used contained fewer than 100,000 entries. Nowadays, the size is increasing exponentially as a result of the genome-sequencing projects, and PMF searches in DNA database banks have to be performed using a six-frame translation of the DNA sequences. As a consequence of the size increase, the confidence level is continually dropping (unless species-specific databases are used), and a higher mass accuracy is required to improve the search accuracy.

6.3.1 Mass Accuracy and Automation

The importance of mass accuracy has been shown by "in silico" (computer-simulated) studies. The increase in confidence reaches a limit with a mass tolerance of 0.01–0.02% (Fenyö et al. 1998). For PMF applications, a fundamental leap in instrumental optimisation was reported in 1995 with the introduction of delayed extraction (Brown and Lennon 1995; Juhasz et al. 1996; Chap. 3). A reflectron MALDI time-of-flight (TOF) can produce measurements routinely with a mass accuracy below 0.05% while maintaining high sensitivity. Combined with a relatively good tolerance for low levels of contaminants, such as buffers and salts, this feature has made MALDI-TOF instruments the

spectrometers of choice for routine fast analysis of unseparated protein digestions. The availability of MALDI targets with hundreds of sample-loading positions, and automatic data acquisition ensures a high sample throughput. In
order to obtain the best spectrum, a balance must be found between the lowest
laser power (which gives the highest resolution) and high laser powers (which
give more sensitivity). Jensen et al. (1997) described the use of fuzzy-logic programming to optimise laser fluence during unattended data accumulation
from MALDI samples. Since the sample is very often unevenly distributed
across a target spot, the spot has to be scanned in order to find a "sweet spot".
However, this is time consuming and may lead to discrimination effects,
because some peptides crystallise with matrix in the centre, while others are
found only at the edges of the spot. In order to circumvent this and to increase
sensitivity, minute sample wells (picovials) produced by etching silica wafers
have been developed, as have as delivery devices (usually ink-drop sprays) that
concentrate the sample into the tiny vials that are almost the size of the laser
beam (Onnerfjord et al. 1998). A single laser shot is often enough to produce
a good spectrum. The effect of improving the mass accuracy of PMF was
described by Vorm and Mann (1994) and, more recently, by Clauser et al.
(1999). When a mass accuracy of ±0.5–5 ppm is achieved, an element composition can be deduced; this drastically reduces the number of matching
peptides. Combining this with the immonium-ion composition of a peptide
obtained by post-source decay or high-energy collision-induced dissociation,
the number of peptide compositions can be reduced so that homology searching becomes possible. Most conservative amino acid replacements involve the
addition or subtraction of a methylene unit (CH_2), such as Gly $\rightarrow$ Ala, Ser $\rightarrow$
Val, Val $\rightarrow$ Lxx, Asp $\rightarrow$ Glu or Asn $\rightarrow$ Gln). If one maintains the mass-accuracy
parameter at approximately 10 ppm, one can allow deviations of approximately
the mass of a methylene group and can perform limited homology searches
(Chap. 9).

6.3.2 Multiple Proteins in a Sample

In proteomics, 2D gels act as the medium connecting the genome sequences
to the level of protein expression in the biological material under investigation. In the "classical" version of this technique, proteins are separated in the
first dimension according to the pI (the pH at which the overall charge on the
protein is zero) and in the second dimension by their MWs. However, the separation power of the method is limited by several factors:

1. High-abundance proteins can mask less-abundant ones
2. Even low-abundance proteins with very close pIs and MWs can co-migrate,
   causing overlapping spots
3. "Trains" of spots due to proteins carrying post-translational modifications
   (like different phosphorylation levels) complicate the gel analysis, increas-

ing the number of detectable proteins and making quantification more difficult

4. Hydrophobic proteins tend to smear or partially precipitate
5. Some contaminants (like cytokeratins) can be introduced during the processing of the sample

As a consequence, multiple proteins are often found in what appear to be single symmetric spots. One can try to circumvent the problem by using narrow pH-range gels (Bjellqvist et al. 1982; Urquhart et al. 1997) to increase the separation in the area of interest, or one can compare different physiological conditions to distinguish between constitutively expressed proteins and induced ones. However, these approaches are tedious and time consuming, and "cross-contamination" of two or more spots cannot be excluded a priori. James et al. (1993) demonstrated the use of PMF to analyse protein mixtures via a search using data from a digest of casein kinase II holoenzyme (and subunits). The two subunits had scores proportional to the number of their peptides used in the search; the scores were approximately proportional to their molar ratios. An example of the identification of a simple protein mixture by PMF alone was shown by Shevchenko et al. (1996a) using a protein contaminated by a cytokeratin. The essential prerequisite for successful analysis was a high mass accuracy of 10–30 ppm.

6.3.3 Protein Modifications

Protein co- and post-translational modifications have been extensively reported in relation to cell development and differentiation; they are often correlated with enzyme activity modulation and targeting. In eukaryotic cells, the use of 2D gels to monitor changes in the glycosylation pattern has characterised tissue specificity and disease states. For example, the train of spots observed for transferrin differs between "normal" persons and those with alcohol-related diseases (Gravel et al. 1996), as does the pattern of spots for serotransferrin in serum and cerebrospinal fluid (Wilkins et al. 1996).

In theory, if the percentage of sequence coverage of a protein is high enough, PMF alone can detect the presence of modifications by observing the mass shifts between the predicted peptides masses and the experimental ones. However, there are several problems with this approach:

1. A high number of modifications makes data interpretation hard and sometimes equivocal
2. The yield of peptides extracted after in-gel or on-membrane digestion can be low, covering just a minor part of the protein
3. Peaks due to contaminants have to be eliminated from the MS spectra
4. Modified peptides do not appear quantitatively with respect to the unmodified, e.g. phosphopeptides are often not observed, because they do not ionise well

The chance that a post-translational modifications will be detected is fairly good when the sequence coverage accounts for at least 80% of the intact protein (Scheler et al. 1998).

6.3.4 Removing Post-Translational Modifications for Searching

Patterson et al. (1995) presented a search with the digestion products of β-casein; the protein could not be identified by fingerprinting, because three out of four peptides carried a modification. In the case of highly modified proteins, such as heavily glycosylated ones, the proteolytic digestion can be difficult to perform. In these situations, the best strategy for protein identification is the elimination of the modification itself. An improved procedure for the chemical deglycosylation of proteins containing sialylated N- and/or O-linked oligosaccharides was described by Raju and Davidson (1994) and was successfully applied to fetuin, tracheal mucin and gastric mucin. The sialic acid residues are first removed by mild acid hydrolysis, then the de-sialylated glycoproteins are treated with anhydrous trifluoromethane sulphonic acid (TFMSA) under conditions that remove all the carbohydrate residues except D-GalNAc linked to serine or threonine. In order to remove the core D-GalNAc residues, the proteins are treated with periodate, followed by a second cycle of TFMSA treatment. For protein de-phosphorylation, a fast and selective procedure has been described by Byford (1991); the β-elimination of phosphoserine and phosphothreonine residues is achieved by incubating the phosphopeptides in 0.1 M Ba(OH)$_2$ solution for approximately 90 min. The reaction is stopped by acidification of the reaction mixture. Compared with other methods of β-elimination by dilute alkalis, the use of Ba(OH)$_2$ catalyses the phosphate release by more than two orders of magnitude without affecting the unmodified serine and threonine residues. Moreover, carboxymethylcysteine is stable under these conditions, and the β-elimination of O-glycosidically linked moieties is not catalysed.

6.3.5 Relative Effectiveness of Different Digesters

In order to test the effectiveness of the mass profiles generated by different cleavage agents in database searches, Wise et al. (1997) performed an "in silico" digestion of the non-redundant protein database. Twenty pseudo-endoproteases cutting at the C-terminus of each amino acid residue and three real enzymes (trypsin, chymotrypsin and endoproteinase GluC) were used. The distribution of the peptide masses obtained was then analysed to determine the "covering set" of a protein, i.e. the subset of fragment masses necessary to distinguish that protein from the others in the database. Keeping a mass accuracy of ±0.5 Da throughout the entire mass range, they found that an average of two peptides was sufficient to identify a protein, though none of the proteases could identify all the proteins in the database. Trypsin, chymotrypsin and GluC were able to cover the largest percentage of entries. The efficiency of protein characterisation was then analysed in relationship to the mass range of the peptides used for the search. A sliding window of 500 Da was used for this experiment, and the low mass range of 350–849 Da exhibited the lowest number of unidentified proteins while masses of over 5000 Da were relatively ineffective. Fenyö et al. (1998) analysed the constraining effects of mass, mass accuracy, the presence of an amino acids and the number of exchangeable hydrogens on the outcome of a search of the *Saccharomyces cerevisiae* genome. They concluded that the higher the mass of a peptide, the more effective it became for protein identification. This appears to contradict the work of Wise; however, the difference is in the size of the database. The non-redundant database contained 128,719 entries against 6129 predicted open reading frames in the yeast genome. As the size of the database increases, the constraining effect of a single peptide decreases. The largest increase in effectiveness of a data set can be obtained if two independent data sets are used.

6.3.6 Practical Limitations of Digestion

There is, unfortunately, a large gap between the "in silico" results and those obtained in practice. The limiting steps in the identification of proteins isolated by 2D gels (and other procedures) are sample handling, protein detection limits on the gel (staining procedure), losses of material during extraction and digestion, digestion itself, and the presence of gel contaminants and/or detergents. These factors become critical once the amount of protein available is 1 pmol or less. The two most commonly used protein-visualisation methods are coomassie blue for protein amounts greater than 10 ng per spot, and silver staining, with a detection threshold of 0.05 ng of protein. Coomassie blue binds to proteins by a combination of hydrophobic and ionic interactions; the latter interfere with tryptic digestion, because the acidic groups pair with lysine and arginine residues, decreasing the digestion efficiency. For a long time, silver staining was believed to be incompatible with further analysis due to the use of Ag^+ as an oxidising agent and the use of glutaraldehyde, a cross-linking agent that blocks free amino groups, as a contrast enhancer. Shevchenko et al. (1996b) showed that, by adopting a simplified procedure with no glutaraldehyde step, even silver-stained spots can be characterised by MS, and little chemical modification takes place. Furthermore, removal of the silver ions by de-staining with ferricyanide and thiosulphate enhances the digestion efficiency (Gharahdaghi et al. 1999).

6.4 Extensions of the Method

6.4.1 Orthogonal Data

One of the main problems of mass mapping lies in assessing the level of confidence that can be placed in the results of protein identification. Studies performed "in silico" or on real samples have both shown that an unambiguous identification can be obtained by using two independent sets of experimental data (orthogonal data; Fig. 6.3). This idea was proposed by James et al. (1994), and the MassSearch program was modified to allow the use of a second set of masses produced by either chemical modification (such as hydrogen-deuterium exchange, methylation or acetylation) of the original digest or from a second digestion using a different protease. A search using a single digestion (especially with only a few peptides) is often not conclusive when performed on a six-frame translation of large DNA databases. Expressed-sequence tags (ESTs), small complementary DNA (cDNA) sequences obtained from random primed cDNA libraries (Adams et al. 1991), are potentially a useful source of information. However, ESTs contain errors, because they are only single-pass sequences, cover only a part of a gene and must always be translated in all six reading frames in order to extract the potential protein sequence. In order to access this data and obtain results with confidence, orthogonal data sets must be used.

Table 6.3 shows the impact of a dual digestion on the final matching list using a protein or DNA database in a search performed with the MassSearch program. Single-data-set searches of the protein database usually produce very clear results ($\Delta > 30$ where Δ is the difference between the top scoring protein and the next non-related protein). When the same data set is run against the DNA database, the confidence level drops drastically, and one cannot be sure

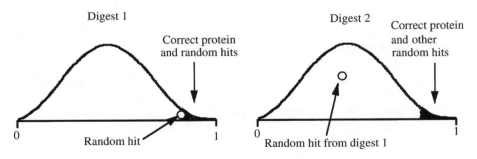

The probability of finding a protein at random follows a normal distribution

Fig. 6.3. Probability and orthogonal data sets. Since there is an element of random noise in a database, the use of two independent data sets can drastically reduce the number of false positive results and increase the confidence level of the search

Table 6.3. The effect of using orthogonal data sets when searching protein and DNA databases. A dual digest involves splitting the protein sample into two aliquots and digesting and measuring each separately. The combined results are used to search the database

Protein	Accession number	Digestions		Protein database single search		DNA database			
		Single	Dual			Single search		Orthogonal search	
				Position	Δ	Position	Δ	Position	Δ
Isocitrate lyase	P05313	Trypsin	AspN	1	80.3	1	8.9	1	228.5
GroESL3	P35862	Trypsin	LysC	1	55.3	1	7.1	1	179.4
50S protein L1	O27716	AspN	Trypsin	1	54.7	2	−2.3	1	189.3
Calmodulin	P02593	CNBr	Trypsin	1	33.9	1	5.5	1	99.7
EF-TU	P02990	GluC	AspN	1	19.4	3	−8.6	1	58.4
Average				1	48.7		2.1	1	151.1

if a top-scoring match really is the protein of interest. In order to restore the confidence level, orthogonal data sets must be used.

6.4.2 Cross-Species Identification

A potential application of the increased number of completely sequenced genomes is the identification of proteins in non-sequenced organisms via cross-species comparison with the sequenced ones. To investigate which parameters can be used in this process (and at what level of confidence), Wilkins and Williams (1997) performed 65 theoretical cross-species comparisons for 21 different proteins: 9 human and 12 from *Escherichia coli*. The amino acid compositions and the MWs of the proteins were generally well conserved among different species, suggesting the use of these parameters for tentative cross-species protein identification. The isoelectric point showed a high degree of variability, while PMF with tryptic digestion could be useful for cross-species matching (but with a low confidence level if the sequence identity was less than 80%, since the fragment profiles were not strictly conserved). A similar study by Cordwell and Humphery-Smith (1997) used 54 ribosomal proteins of *Mycoplasma genitalium* and 72 aminoacyl transfer RNA synthases from *Haemophilus influenzae*, *M. genitalium* and *Methanococcus jannaschii* genomes, which gave a wide range of sequence identities (22.7–100%). A higher percentage of cross-species protein identification was achieved when using a special program (COMBINED) that combines the score list based on the amino acid composition with the score list obtained by PMF (orthogonal data processing). A wide mass tolerance (±6 Da) was suggested by the authors to compensate for the small percentage of amino acid substitutions. An alternative/complementary approach could use dual digestions as orthogonal data. Interestingly, to constrain the identification, the predicted length of a sequence tag from a fragment was calculated to be between 11 and 20 amino acids, thus limiting – theoretically – this approach for protein characterisation among different species (Cordwell and Humphery-Smith 1997). A novel aspect of PMF is the identification of conserved peptide masses among species. The selection of one or more conserved motifs as fragment masses could provide stringent and effective tools for the fast recognition of homologous proteins among different species (Cordwell et al. 1997). These conserved motifs should reflect the presence of conserved functional domains whose integrity was preserved during evolution.

6.4.3 Chemical Modifications

Hydrogen–deuterium exchange is an established derivatisation method in mass analysis and has been shown to be an effective aid in the interpretation of MS/MS spectra for peptide sequencing (Sepetov et al. 1993; Spengler et al. 1993). Spengler et al. developed a method for the deuteration of peptides applied to the target of a MALDI mass spectrometer. The sample is mixed with the matrix, applied to the target and analysed in an unmodified condition. Deuteration of the sample is then performed in a special vacuum chamber connected to a rotary pump and a dry nitrogen supply. Once the target is placed in the chamber, the system is flushed with dry nitrogen, a 1:1 mixture of D_2O and C_2H_5OD is applied to the top of the sample with a syringe and evaporated under a vacuum. The procedure is repeated, then the target is placed back in the mass spectrometer for a second analysis. The efficiency of the procedure is approximately 98%; thus, the spectra of deuterated samples basically exhibit only the signals of fully deuterated ions, shifted in mass by the number of exchangeable hydrogens. The number of exchangeable hydrogens on a peptide depends on the sequence, so peptides with similar MWs can be distinguished after exchange (A, F, G, I, L, M and V have one exchangeable hydrogen; C, D, E, H, S, T, W and Y have two; K, N, carboxymethylcysteine and Q have three; R has five and P none).

An example of a search using tryptic digestion of creatine kinase B from chicken gizzards is shown in Table 6.4. Although PMF only placed the correct protein at position three, the deuterium exchange allowed the elimination of the other potential matches from the list. Other chemical modifications are used in peptide sequencing by MS, such as acetylation (which gives the number of lysines in a peptide and is most effective in non-tryptic or LysC digestion) and methylation (which gives the number of acidic residues and is most effective in non-GluC or AspN digestion).

6.4.4 Iterative Searching

To simplify the identification of protein mixtures present in single gel spot, Jensen et al. (1996) suggested an iterative approach. A MALDI-MS map was

Table 6.4. Deuterium-exchange data and peptide-mass fingerprinting. Creatine kinase B was digested with trypsin, and the following masses were used to search the protein database: 575, 685, 691, 759, 1133 and 1232. Since the number of peptides and their accuracies were relatively low, creatine kinase was only ranked third. The tryptic digest was re-dissolved in deuterium oxide (D_2O), and the following masses were obtained: 585, 698, 703, 772, 1150 and 1251. By comparing the expected and observed mass shifts, creatine kinase could be identified with a high degree of confidence from among the 100 highest-scoring proteins. Upon deuteration, the expected mass shifts for each amino acid are: A = 1, C = 2, D = 2, E = 2, F = 1, G = 1, H = 2, I = 1, K = 3, L = 1, M = 1, N = 3, P = 0, Q = 3, R = 5, S = 2, T = 2, V = 1, W = 2 and Y = 2

Position	Protein	Matching masses	Mass increase after deuteration	Calculated mass–observed mass
1	Ubiquinone-binding protein	691	14	2
		759	14	1
		1134	21	4
		1231	21	3
2	Proline-specific permease	688	17	4
		759	10	3
		1134	27	10
		1231	19	0
3	Creatine kinase B	685	13	0
		692	11	1
		759	13	0
		1232	19	0
4	Patatin-B1 precursor	575	13	3
		759	13	0
		1133	19	2
5	Patatin-B2 precursor	575	13	3
		759	13	0
		1133	19	2

generated for trypsin-digested protein spots and, using the PeptideSearch program, an initial database search was performed to identify a candidate as the most prominent component. The highest-scoring protein is then compared with the measured peptide masses and is checked for other masses arising from incomplete cleavage and peptide modifications (such as oxidised methionine or S-acrylamidocysteine). The list of remaining unassigned peptide masses is then used for a second-pass database search. The procedure is then repeated until no significant match is found. The authors suggested "filtering" the MS mapping data, by eliminating the masses of human keratins (operator fingerprinting), thus "cleaning" the MS spectrum and reducing the number of false-positive matches in the search. Moreover, the mass accuracy should be kept rather high (30 ppm). When one protein is present in large excess, the detection of rare components is ineffective, and the alternative strategy of MS/MS sequencing is preferable.

The problem of filtering experimental data has been examined by Parker et al. (1998). The authors analysed 14 2D gel spots (from a yeast extract) which remained unidentified after a PMF search performed at high accuracy. The aim was to recognise and identify peptides arising from cross-contaminating proteins and to subtract these masses from subsequent MS spectra. The first step was to construct a source protein database containing all the proteins expected to be present (trypsin, keratins, previously identified abundant proteins). The computer program is then able to inform the operator if some of the masses from the MS spectra of unidentified spots have been encountered before. Masses commonly retrieved in spectra from different spots are likely to be due to an external contaminant or to a protein that is heterogeneous and widely distributed in the gel.

6.5 Sequence Tagging

Incomplete protein cleavage often occurs, resulting in poor protein-sequence coverage for low-abundance or hydrophobic proteins, and subsequent PMF gives low confidence results. A powerful extension of database searching with MS data is the addition of a small amount of peptide-sequence data as a search parameter. Two different approaches have been proposed. The MS/MS fragmentation spectrum can be used to directly search the sequence database in a fully automated fashion (Eng et al. 1994) and is described in detail in Chapter 7. Alternatively, the spectrum can be manually inspected in order to read a partial sequence or tag for searching (Mann and Wilm 1994) and is described in Chapter 8.

6.5.1 Ragged Termini

Alternative approaches have been developed to obtain sequence tags without performing MS peptide-fragmentation experiments. Jensen et al. (1996) observed that, in tryptic digestions, pairs of peptides separated by the molecular mass of a lysine or arginine could be shown on a MALDI spectrum due to the incomplete cuts at adjacent cleavage sites. The analysis of a non-redundant protein-sequence database revealed that a protein with a mass greater than 20 kDa contains (on average) four or more sequence combinations of the types KK, RK, KR and RR. Since most endoproteases are poor exopeptidases, these will generate ragged termini, because the chances are 50:50 that cutting will take place at the first rather than the second residue. The use of this single N- or C-terminal amino acid tag increases the search specificity by a factor of 20. The generation of ragged termini was also reported by Tsugita et al. (1998) after chemical digestion of a protein. The authors observed that, in the case of AspC cleavage performed with pentafluoropropionic acid vapour, more than half of the produced fragments exhibited the release of the exposed C-terminal aspartic acid, giving rise to pairs of peaks separated by a mass of

115 Da on the MS spectrum. The group of Peter Roepstorff (Gobom et al. 1999) subsequently confirmed this.

6.5.2 Enzymatic-Ladder Sequencing

Several groups have reported the use of carboxy- and amino-peptidases to generate a ladder of degradation products. The mass difference between the adjacent peaks defines the amino acid being removed (and, thus, the peptide sequence). The development of MALDI-TOF MS has greatly simplified the use of the method. The sample is applied to the target in a series of positions; a series of dilutions of the enzyme are then added for digestion. The digests obtained for the various positions are then combined to reconstruct the sequence ladder (Patterson et al. 1995). Recently, we have developed an algorithm (MassDynSearch) for protein identification using a combination of peptide masses and small associated sequences (tags) generated enzymatically or chemically (Korostensky et al. 1998). The protein of interest is digested, and the resultant fragments are subjected to partial exopeptidase degradation (Fig. 6.4).

The MALDI-TOF spectra of the digestion before and after the degradation give a list of intact peptide masses, each associated with a set of degradation products. The MassDynSearch algorithm uses these "tagged masses" to search for proteins with similar tagged motifs in protein or DNA databases. The main advantage of this approach is that less-specific proteases (chymotrypsin, elastase, pepsin, etc.) can also be used. This is especially useful for proteins that are very difficult to fragment with the specific proteases commonly adopted for PMF and for very insoluble proteins (these can be dissolved in 40% formic acid and digested with pepsin prior to exopeptidase digestion). The main drawback is the relatively large amount of material required (many femtomoles) and the unpredictable activity of the exopeptidase towards the various peptides, making the extraction of ladder sequences from unseparated protein digests very difficult.

6.5.3 Chemical-Ladder Sequencing

Chait et al. (1993) suggested the use of a modified version of Edman degradation to generate a set of sequence-defining peptide fragments. The degradation is performed using phenylisothiocyanate with a small amount of a chain terminator (phenylisocyanate), which produces a small portion (<5%) of blocked peptide chain during each cycle. After a predetermined number of cycles, the resulting mixture is analysed with a MALDI-MS instrument and the sequence is determined from the mass differences between the termination products. The main disadvantage is that the reactions are performed with the sample immobilised on a polymeric membrane. Small peptides and low

Incomplete protein digest

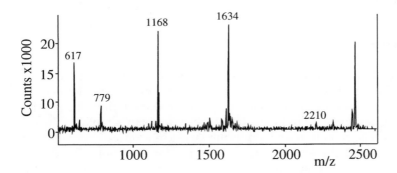

Carboxypeptidase treatment of digest

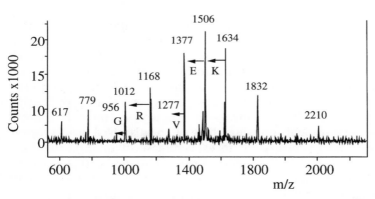

Fig. 6.4. Enzyme-ladder sequence tag. *Above* A matrix-assisted laser desorption ionisation time-of-flight spectrum of a tryptic digest of 10 pmol of cytochrome C. *Below* The same digest was then incubated on a target with carboypeptidase P and re-measured. Two peptides were digested, giving mass tags of 1634.9 (K/Q, E, V) and 1168 (R, G), which were used by the MassDynSearch algorithm to positively identify the protein as cytochrome C

amounts of sample suffer from extensive losses during washing steps with organic solvents, when the excess phenylisothiocyanate is removed. Conversely, peptides that bind well are difficult to remove for the final MS analysis. A modification of this method in which a volatile fluoroisothiocyanate is used to avoid losses by washing was described by Bartlet-Jones et al. (1994). The sample is divided into aliquots, depending on the number of cycles to be performed; after each cycle, another aliquot is added. The problems with this procedure are the number of handling steps and the difficulty in creating aliquots of low quantities of peptides (as often occurs in proteome studies) without major losses. Gu and Prestwich (1997) suggested the replacement of the volatile amino-modification reagent with allylisothiocyanate.

Recently, another procedure was developed in our lab (Hoving et al. 2000) using a water-soluble degradation reagent. In order to reduce sample handling and loss, short ladder sequences are generated using aqueous solutions via thioacetylation under basic conditions where the N-termini of peptides are immobilised on reverse-phase material. Subsequent cleavage of the N-termini of the thioacetylated peptides is performed by vapour-phase hydrolysis with trifluoroacetic acid. All washing steps are performed using aqueous solutions, so there is no sample washout, and recovery is easy with organic solvents. The method can be applied to pure peptides or mixtures of them (Fig. 6.5), and a prototype machine enabling full automation has been described.

6.6 Conclusion

PMF is a well-established, essential starting procedure for the rapid identification of proteins in proteome projects. It has often been used as the sole means of protein identification, as in the case of an enriched preparation of spindle poles of *Saccharomyces* (where 12 isolated proteins were identified by MALDI peptide-mass mapping alone; Wigge et al. 1998). With the extensions to the method described in Sections 6.4 and 6.5, the method can theoretically be successfully applied to every protein contained in a database. In combination with subtractive 2D-gel analysis, it is the quickest screening system for the study of complex processes (for example, the various steps involved in the processing of a prohormone; Li et al. 1994). Another good example is the development of a comprehensive 2D database of proteins from the urine of patients with bladder cancer (Rasmussen et al. 1996) in order to identify tumour-associated proteins for use as diagnostic/prognostic factors. Similarly, a major effort is being directed at characterising specific banks of peptides. For example, there is a vast number of potentially clinically important peptides present in human haemofiltrate, which is available in large amounts from patients suffering from chronic renal failure (Schulz-Knappe et al. 1997).

The challenge for the future is the full automation of all the steps involved in this approach, using either 2D chromatographic methods for peptide banks (Opiteck et al. 1998) or robotic systems for the excision of spots from 2D gels (Traini et al. 1998), the subsequent digestion of the proteins and their analysis by MS. The major hurdle to overcome is sample handling at increasingly low levels because, although MS instruments are reaching low-attomole sensitivities, protein digestion and handling below the 100-fmol level is not routinely possible.

(1) Tryptic digest

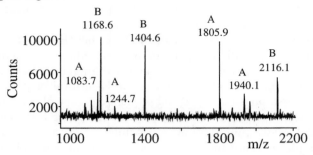

(2) Chymotryptic digest

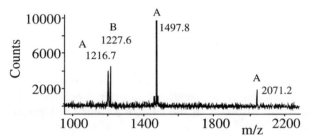

(3) Chymotryptic digest after tagging

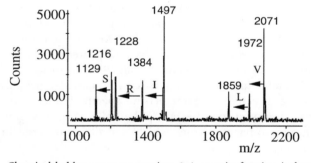

Fig. 6.5. Chemical-ladder sequence tagging. **1** A protein fraction isolated from ribosomes of *Escherichia coli* was digested with trypsin, and the matrix-assisted laser desorption ionisation spectrum was measured. Peptide-mass fingerprinting gave no positive identification of the protein(s). **2** A second aliquot of the protein fraction was digested with chymotrypsin, and the masses were used alone and in combination with the tryptic masses for fingerprint searching. No positive identification was obtained. **3** The same chymotryptic digest was then subjected to three rounds of partial thioacetylation cleavage, which generated two peptide tags: 2071 (*V*, *L*) and 1497 (*I*, *R*, *S*). MassDynSearch identified the protein as EC-3227 (rplX; 50S ribosomal subunit protein L24). The tryptic masses corresponding to this protein are marked with an *A* in **1**. The remaining masses were used to search the database again, and a second protein, EC-3225 (rpsN; 30S ribosomal subunit protein S14) could be identified. The corresponding tryptic fragments are labelled *B* in **1**

References

Adams MD, Kelley JM, Gocayne JD, Dubnick M, Polymeropoulos MH, Xiao H, Merril CR, Wu A, Olde B, Moreno RF, Kerlavage AR, McCombie WR, Venter JC (1991) Complementary DNA sequencing: expressed sequence tags and human genome project. Science 252:1651–1656

Bartlet-Jones M, Jeffery WA, Hansen HF, Pappin, DJ (1994) Peptide ladder sequencing by mass spectrometry using a novel, volatile degradation reagent. Rapid Commun Mass Spectrom 8:737–742

Bjellqvist B, Ek K, Righetti PG, Gianazza E, Gorg A, Westermeier R. Postel W (1982) Isoelectric focusing in immobilized pH gradients: principle, methodology and some applications. J Biochem Biophys Methods 6:317–339

Brown RS, Lennon JJ (1995) Mass resolution improvement by incorporation of pulsed ion extraction in a matrix-assisted laser desorption/ionization linear time-of-flight mass spectrometer. Anal Chem 67:1998–2003

Byford MF (1991) Rapid and selective modification of phosphoserine residues catalysed by Ba2+ ions for their detection during peptide microsequencing. Biochem J 280:261–265

Chait BT, Wang R, Beavis RC, Kent SB (1993) Protein ladder sequencing. Science 262:89–92

Clauser KR, Baker P, Burlingame AL (1999) Role of accurate mass measurement in protein identification strategies employing MS or MS/MS and database searching. Anal Chem 71:2871–2882

Cleveland DW, Fischer SG, Kirschner MW, Laemmli UK (1977) Peptide mapping by limited proteolysis in sodium dodecyl sulfate and analysis by gel electrophoresis. J Biol Chem 252:1102–1106

Cordwell SJ, Humphery-Smith I (1997) Evaluation of algorithms used for cross-species proteome characterisation. Electrophoresis 18:1410–1417

Cordwell SJ, Wasinger VC, Cerpa-Poljak A, Duncan MW, Humphery-Smith I (1997) Conserved motifs as the basis for recognition of homologous proteins across species boundaries using peptide-mass fingerprinting. J Mass Spectrom 32:370–378

Eng JK, McCormack AL, Yates JR III (1994) Correlating tandem mass spectral data of peptides to sequences in a protein database. J Am Soc Mass Spectrom 5:976–989

Fenyö D, Qin J, Chait BT (1998) Protein identification using mass spectrometric information. Electrophoresis 19:998–1005

Gharahdaghi F, Weinberg CR, Meagher DA, Imai BS, Mische SM (1999) Mass spectrometric identification of proteins from silver-stained polyacrylamide gel: a method for the removal of silver ions to enhance sensitivity. Electrophoresis 20:601–605

Gibson BW, Biemann K (1984) Strategy for the mass spectrometric verification and correction of the primary structures of proteins deduced from their DNA sequences. Proc Natl Acad Sci USA 81:1956–1960

Gobom J, Mirgorodskaya E, Nordhoff E, Hojrup P, Roepstorff P (1999) Use of vapor-phase acid hydrolysis for mass spectrometric peptide mapping and protein identification. Anal Chem 71:919–927

Gravel P, Walzer C, Aubry C, Balant LP, Yersin B, Hochstrasser DF, Guimon J (1996) New alterations of serum glycoproteins in alcoholic and cirrhotic patients revealed by high resolution two-dimensional gel electrophoresis. Biochem Biophys Res Commun 220:78–85

Greer FM, Morris HR, Forstrom J, Lyons D (1988) Fast atom bombardment mass spectral search for the amino terminus of genetically engineered alpha 1-antitrypsin. Biomed Environ Mass Spectrom 16:191–195

Gu QM, Prestwich GD (1997) Efficient peptide ladder sequencing by MALDI-TOF mass spectrometry using allyl isothiocyanate. J Pept Res 49:484–491

Henzel WJ, Billeci TM, Stults JT, Wong SC, Grimley C, Watanabe, C (1993) Identifying proteins from two-dimensional gels by molecular mass searching of peptide fragments in protein sequence databases. Proc Natl Acad Sci USA 90:5011–5015

Hoving S, Münchbach M, Schmid H, Signor L, Lehmann A, Staudenmann W, Quadroni M, James P (2000) A method for the chemical generation of N-terminal peptide sequence tags for rapid protein identification. Anal Chem 72: 1006–1014

James P, Quadroni M, Carafoli E, Gonnet G (1993) Protein identification by mass profile fingerprinting. Biochem Biophys Res Commun 195:58–64

James P, Quadroni M, Carafoli E, Gonnet G (1994) Protein identification in DNA databases by peptide mass fingerprinting. Protein Sci 3:1347–1350

Jensen ON, Vorm O, Mann M (1996) Sequence patterns produced by incomplete enzymatic digestion or one-step Edman degradation of peptide mixtures as probes for protein database searches. Electrophoresis 17:938–944

Jensen ON, Mortensen P, Vorm O, Mann M (1997) Automation of matrix-assisted laser desorption/ionization mass spectrometry using fuzzy logic feedback control. Anal Chem 69:1706–1714

Juhasz P, Roskey MT, Smirnov IP, Haff LA, Vestal ML, Martin SA (1996) Applications of delayed extraction matrix-assisted laser desorption ionization time-of-flight mass spectrometry to oligonucleotide analysis. Anal Chem 68:941–946

Korostensky C, Staudenmann W, Dainese P, Hoving S, Gonnet G, James P (1998) An algorithm for the identification of proteins using peptides with ragged N- or C-termini generated by sequential endo- and exopeptidase digestions. Electrophoresis 19:1933–1940

Li KW, Hoek RM, Smith F, Jimenez CR, van der Schors RC, van Veelen PA, Chen S, van der Greef J, Parish DC, Benjamin PR, Jeraerts W (1994) Direct peptide profiling by mass spectrometry of single identified neurons reveals complex neuropeptide-processing pattern. J Biol Chem 269:30288–30292

Mann M, Wilm M (1994) Error-tolerant identification of peptides in sequence databases by peptide sequence tags. Anal Chem 66:4390–4399

Mann M, Hojrup P, Roepstorff P (1993) Use of mass spectrometric molecular weight information to identify proteins in sequence databases. Biol Mass Spectrom 22:338–345

Morris HR, Panico M, Etienne T, Tippins J, Girgis SI, MacIntyre I (1984) Isolation and characterization of human calcitonin gene-related peptide. Nature 308:746–748

Onnerfjord, P, Nilsson J, Wallman L, Laurell T, Marko-Varga G (1998) Picoliter sample preparation in MALDI-TOF MS using a micromachined silicon flow-through dispenser. Anal Chem 70:4755–4760

Opiteck GJ, Ramirez SM, Jorgenson JW, Moseley MA III (1998) Comprehensive two-dimensional high-performance liquid chromatography for the isolation of overexpressed proteins and proteome mapping. Anal Biochem 258:349–361

Pappin DJC, Hojrup P, Bleasby AJ (1993) Rapid identification of proteins by peptide-mass fingerprinting. Curr Biol 3:327–332

Parker KC, Garrels JI, Hines W, Butler EM, McKee AH, Patterson D, Martin S (1998) Identification of yeast proteins from two-dimensional gels: working out spot cross-contamination. Electrophoresis 19:1920–1932

Patterson DH, Tarr GE, Regnier F, Martin SA (1995) C-terminal ladder sequencing via matrix-assisted laser desorption mass spectrometry coupled with carboxypeptidase Y time-dependent and concentration-dependent digestions. Anal Chem 67:3971–3978

Patterson SD, Aebersold R (1995) Mass spectrometric approaches for the identification of gel-separated proteins. Electrophoresis 16:1791–1814

Raju TS, Davidson EA (1994) New approach towards deglycosylation of sialoglycoproteins and mucins. Biochem Mol Biol Int 34:943–954

Rasmussen HH, Orntoft TF, Wolf H, Celis JE (1996) Towards a comprehensive database of proteins from the urine of patients with bladder cancer. J Urol 155:2113–2139

Scheler C, Lamer S, Pan Z, Li XP, Salnikow J, Jungblut P (1998) Peptide mass fingerprint sequence coverage from differently stained proteins on two-dimensional electrophoresis patterns by matrix assisted laser desorption/ionization-mass spectrometry. Electrophoresis 19:918–927

Schulz-Knappe P, Schrader M, Standker L, Richter R, Hess R, Jurgens M Forssmann W (1997) Peptide bank generated by large-scale preparation of circulating human peptides. J Chromatogr A 776:125–132

Sepetov NF, Issakova OL, Lebl M, Swiderek K, Stahl DC, Lee TD (1993) The use of hydrogen-deuterium exchange to facilitate peptide sequencing by electrospray tandem mass spectrometry. Rapid Commun Mass Spectrom 7:58–62

Shevchenko A, Jensen ON, Podtelejnikov AV, Sagliocco F, Wilm M, Vorm O, Mortensen P, Boucherie H, Mann M (1996a) Linking genome and proteome by mass spectrometry: large-scale identification of yeast proteins from two dimensional gels. Proc Natl Acad Sci USA 93:14440–14445

Shevchenko A, Wilm M, Vorm O, Mann M (1996b) Mass spectrometric sequencing of proteins silver-stained polyacrylamide gels. Anal Chem 68:850–858

Spengler B, Lutzenkirchen F, Kaufmann R (1993) On-target deuteration for peptide sequencing by laser mass spectrometry. Org Mass Spectrom 28:1482–1490

Traini M, Gooley AA, Ou K, Wilkins MR, Tonella L, Sanchez JC, Hochstrasser DF, Williams KL (1998) Towards an automated approach for protein identification in proteome projects. Electrophoresis 19:1941–1949

Tsugita A, Kamo M, Miyazaki K, Takayama M, Kawakami T, Shen R, Nozawa T (1998) Additional possible tools for identification of proteins on one- or two- dimensional electrophoresis. Electrophoresis 19:928–938

Urquhart BL, Atsalos TE, Roach D, Basseal DJ, Bjellqvist B, Britton W, Humphery-Smith I (1997) 'Proteomic contigs' of *Mycobacterium tuberculosis* and *Mycobacterium bovis* (BCG) using novel immobilised pH gradients. Electrophoresis 18:1384–1392

Vorm O, Mann M (1994) Improved mass accuracy in matrix assisted laser desorption ionisation time of flight mass spectrometry of peptides. J Am Soc Mass Spectrom 5:955–958

Wigge PA, Jensen ON, Holmes S, Soues S, Mann M, Kilmartin J (1998) Analysis of the Saccharomyces spindle pole by matrix-assisted laser desorption/ionization (MALDI) mass spectrometry. J Cell Biol 141:967–977

Wilkins MR, Williams KL (1997) Cross-species protein identification using amino acid composition, peptide mass fingerprinting, isoelectric point and molecular mass: a theoretical evaluation. J Theor Biol 186:7–15

Wilkins MR, Sanchez JC, Williams KL, Hochstrasser DF (1996) Current challenges and future applications for protein maps and post-translational vector maps in proteome projects. Electrophoresis 17:830–838

Wise MJ, Littlejohn TG, Humphery-Smith I (1997) Peptide-mass fingerprinting and the ideal covering set for protein characterisation. Electrophoresis 18:1399–1409

Yates JR III, Speicher S, Griffin PR, Hunkapiller T (1993) Peptide mass maps: a highly informative approach to protein identification. Anal Biochem 214:397–408

7 Protein Identification by SEQUEST

David L. Tabb, Jimmy K. Eng, and John R. Yates III

7.1 Introduction to SEQUEST and Protein Identification

Sequencing peptides was an enormously cumbersome process until the development of tandem mass spectrometry (MS/MS). The main barrier to broad use of this technique has been the difficulty of spectrum interpretation. The SEQUEST software package attempts to ease this process by automatically matching tandem mass spectra to database sequences (Eng et al. 1994). Candidate sequences are culled from the database by several simple filters. Virtual spectra are constructed for these sequences, and these are compared to the observed spectrum. Sequences yielding spectra most similar to the observed one are reported to the user.

Analysis of similarity scores can give a measure of the certainty of identifications. For an individual spectrum, cross-correlation scores provide information about the quality of a match between a sequence and observed fragmentation peaks. For a collection of spectra from the same sample, several peptides may be matched to a single protein, providing greater certainty of its presence. Comparative analysis can highlight proteins related to the differences among samples. A key to the effective use of SEQUEST is an understanding of its output.

The utility of this software has been extended in several ways. SEQUEST and its helper programs provide multiple methods for identifying peptides containing post-translational modifications. In addition, SEQUEST can recognise and disregard spectra that are excessively noisy or that represent contaminants. Running the software on multiple machines simultaneously can enhance its performance. These extensions make the software more flexible and powerful. SEQUEST has been applied to peptide spectra from many types of purification. It is particularly powerful in cataloguing components of complex mixtures, and it can be used to pinpoint peptide spectra from proteins that are not in the database. Its use is not limited to any particular isolation protocol; SEQUEST can be used with two-dimensional gel separations or with chromatographic or centrifugal ones.

7.1.1 Peptide MS and the Collision-Induced Dissociation Process

Most mass-spectrometric sequencing strategies analyse peptides rather than full-length proteins. Several techniques exist for introducing peptides into a mass spectrometer. Proteins are commonly digested by trypsin, because the resultant peptides produce spectra that are more amenable to interpretation. To enable the mass spectrometer to analyse the peptides, gas-phase ions must be formed through electrospray ionisation or matrix-assisted laser desorption ionisation (Chaps. 2, 3). A mass analyser then scans through the m/z (ratio of mass to charge) range. A detector records the ions, producing a mass spectrum.

An ion of interest can then be selected for fragmentation. Energy is supplied to the selected peptide ion through collisions with an inert gas or a surface. This energy causes the peptide ions to fragment at different points. The recorded product ions comprise the tandem mass spectrum.

Peptide fragmentation is not entirely random; the pattern of breakage is dependent upon the parameters of the collision-induced dissociation and other sequence-specific features. The breaks commonly occur at the peptide bonds, resulting in A, B and Y ion types (Fig. 7.1; Chap. 8). B ions are formed when the charge of the peptide is on the N-terminal side, while Y ions are formed when the charge stays on the C-terminal side. The B ion including only the first residue of the peptide is called B1, and the remaining B ions are numbered progressively higher as one progresses toward the C-terminus (Fig. 7.2). Y ions are numbered similarly but start at the C-terminus. Other ions can either form directly or result from the secondary fragmentation of major fragment ions. An understanding of the fragmentation process underlies the strategies for peptide sequencing through MS/MS.

7.1.2 Interpretation of Tandem Mass Spectra

Manually inferring a peptide sequence from a tandem mass spectrum begins with identifying a series of ions. For example, if one can identify a string of intense peaks in the high-mass range as Y ions, the differences in mass between

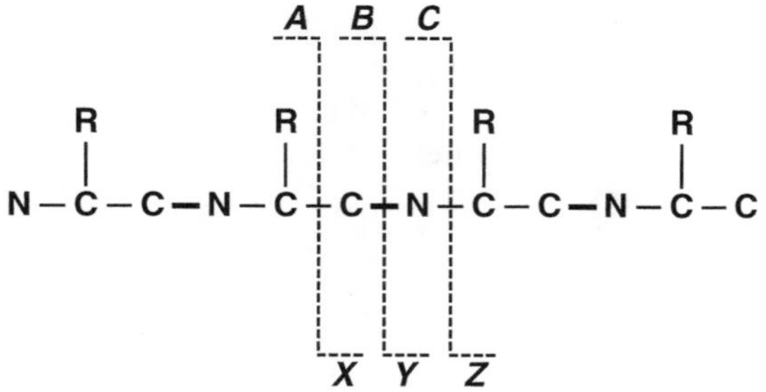

Fig. 7.1. Fragments resulting from backbone cleavages. B and Y ions are the most common

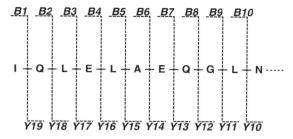

Fig. 7.2. Fragment numbering for an example sequence

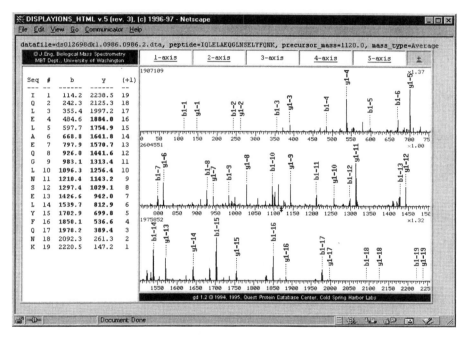

Fig. 7.3. A tandem mass spectrum for the example sequence in SEQUEST's web-browser interface

adjacent peaks should correspond to the masses of amino acid residues (Fig. 7.3). Serially subtracting masses of adjacent peaks can provide a sequence inference that spans an entire spectrum. Unfortunately, some expected peaks can be missing. In addition, there are often several possible inferences for each peak, and it is not always clear which represents the correct residue.

Many challenges are posed by manual sequencing. Spectra are frequently complicated by ions that are fragmentation products of fragmentation products. Certain peptide residues can bias fragmentation, reducing the abundance of some fragments significantly. Another problem plaguing manual interpretation is that identifying the fragmentation series from which an ion comes is not always possible. Manual sequencing requires a clear spectrum and a patient investigator.

7.2 Description of the Algorithm

SEQUEST handles the sequence-inference problem very differently. Instead of attempting to interpret the spectrum directly, it identifies which sequence from a database matches the spectrum best. Manual interpretation may produce several sequences to be tested against databases for confirmation. SEQUEST, however, produces virtual spectra from database sequences to test against the observed spectrum. SEQUEST's comparison occurs in the spectrum domain, while manual comparisons occur in the sequence domain.

The algorithm proceeds through several stages. In the first, spectrum pre-processing, the software simplifies the observed spectrum to make it more amenable to rapid comparison. Next, it selects peptide sequences that have the same m/z ratio (within a defined range) as the observed precursor from a database. These sequences are further filtered by simple attempts to match them to the spectrum. For each of the remaining spectra, SEQUEST creates a "virtual" spectrum and cross-correlates it with the observed spectrum. The sequences are ranked by these scores and reported.

7.2.1 Spectrum Pre-Processing

The observed spectrum can be simplified for improved performance and accuracy. Multiple scans of the fragment ions are combined into a single scan, and the m/z value of each peak is rounded to the nearest integer. Since most types of tandem mass spectra contain intact precursor ions, this peak and a small area around it are removed from the spectrum. This removal prevents the precursor from later being misidentified as a fragment ion. A quota is imposed on the remaining peaks; only the top 200 ions are considered during sequence-candidate selection, ignoring many noise peaks and simplifying analysis. The resulting spectrum is simpler, enriched for putative fragment ions and missing many peaks that might otherwise complicate processing or produce false positive results.

7.2.2 Selection of Candidate Sequences from a Database

SEQUEST next selects peptides from a database of protein sequences. The first selection is based on the known precursor-ion mass. If the algorithm has been instructed not to bias for a particular enzyme cleavage site, it will attempt to find any peptide at or near the mass of the parent ion that was subjected to fragmentation. If the researcher has used trypsin, for example, the algorithm performance can be improved by including only those peptides that would be formed by complete or incomplete tryptic digestion. Thus, the precursor-ion mass is the first information used in identifying plausible sequence matches.

Each peptide sequence in the set is compared with the observed spectrum. Each sequence is converted into a list of m/z values at which fragment ions would occur. The spectrum is searched for ions at these locations, and the intensity of these ions is summed. The continuity of each sequence is evaluated by providing a bonus for successive fragment ions. The percentage of ions found versus those expected is also calculated. These three factors are combined to produce a score representing a quick analysis of each sequence in comparison with the spectrum.

7.2.3 Preparation for Spectrum Comparison

Techniques exist with which to compare sequences to other sequences effectively, and comparing spectra to other spectra is also well studied; however, bridging the two domains is more complex. SEQUEST incorporates a method to compare a sequence to a spectrum by constructing a virtual tandem mass spectrum from each of the 500 best-scoring sequences in the preliminary round of evaluation. SEQUEST calculates the m/z values for each of the ions predicted to result from cleavage at each peptide bond and constructs a spectrum with peaks at these locations. Ion abundance can be independently adjusted for each fragment-ion type. The virtual spectrum is then compared with the experimental tandem mass spectrum.

The observed spectrum is normalised further in preparation for comparison. It is divided into ten equal regions according to the m/z. The tallest peak in each region is assigned a height of 50.0, and the heights of all other peaks in the region are normalised with respect to this peak. This normalisation removes the variation in intensity that occurs as a function of m/z.

7.2.4 Cross-Correlation

Each virtual spectrum is compared with the normalised spectrum to produce a cross-correlation score. Fast Fourier transforms of the observed and the virtual spectra are manipulated to produce a correlation score. The virtual spectrum with the highest score indicates the sequence that best matches the observed spectrum.

A signal with many peaks has a better chance of aligning with another signal, so relying strictly on simple alignments would bias an algorithm toward more complex virtual spectra and, thus, longer sequences. SEQUEST relies on the difference between the alignment of the two spectra with no offset and the average alignment when the two spectra are offset from each other by small values (Powell and Heiftje 1978). This difference rises with the similarity between the observed spectrum and the virtual spectrum and does not suffer from the bias.

A brief summary of the SEQUEST algorithm includes the following steps: SEQUEST searches the database for candidate peptides on the basis of mass alone. It creates a virtual spectrum for each of these peptides and checks to see how well it matches the observed spectrum. The peptides matching the spectrum best are listed with each of their scores. At the expense of some computer time, a large batch of tandem mass spectra can be assigned to peptide sequences from a database.

7.3 Interpretation of SEQUEST Scores

When SEQUEST searches a sequence database to produce a list of plausible peptide matches, it produces a list of possible answers. The most naive interpretation of the results is to simply assume that the top answer is correct, but a deeper understanding of score interpretation can lead to more accurate assessment of the search results. Careful consideration of the scores can yield the most information from each spectrum, from collections of spectra resulting from a single liquid chromatography (LC) MS/MS run and even from comparisons between multiple LC/MS/MS runs.

7.3.1 Interpretation of a Spectrum's SEQUEST Results

The SEQUEST output file includes several variables that describe the correspondence of each spectrum to the listed sequences. An example of an output file is shown in Fig. 7.4. The header contains information about the search, including the date and time the search was executed, the name of the queried database, the database statistics, and the search parameters specified for the analysis (including mass tolerance), etc. The peptides best matched in the analysis are listed from best to worst below the header. The columns display the theoretical protonated peptide mass, preliminary and correlation scores, rankings, and identified proteins from which the peptide may have come. The bottom of the output file displays the full protein reference text for a specified number of the identified peptides.

An understanding of the output file leads to better analysis of the search results. Determining the reliability of the top-ranked sequence match is usually the first priority. At the most basic level, this requires a user to compare the expected theoretical fragment ions for the peptide to the peaks in the input spectrum. This process is facilitated by SEQUEST's web-browser interface. A click of a mouse button shows a display of the input spectrum, with labels at locations predicted to be fragments of the highest-ranking sequence (Fig. 7.3).

```
ds012698dk1.0986.0986.2.out

SEQUEST v.27 (rev. 1.5), (c) 1993

Molecular Biotechnology, Univ. of Washington, J.Eng/J.Yates

Licensed to John Yates III @ University of Washington

03/17/1999, 03:20 PM, 10 sec. on squatch

(M+H)+ mass = 2239.0700 ~ 3.0000 (+2), fragment tol = 0.0, AVG/AVG

total inten = 10074.0, lowest Sp = 280.7, # matched peptides = 272951

# amino acids = 5332534, # proteins = 11506, /wfs/dbase/nci/yeast.nci

ion series nABY ABCDVWXYZ: 0 1 1 0.0 1.0 0.0 0.0 0.0 0.0 0.0 1.0 0.0

display top 5/3, ion % = 0.0, CODE = 0000
```

#	Rank/Sp	(M+H)+	deltCn	XCorr	Sp	Ions	Reference		Peptide
1.	1 / 1	2238.5022	0.0000	6.3045	2912.9	26/36	SW:YAD5_YEAST	+1	K.IQLELAEQGLNSELYFQNK.N
2.	2 / 7	2237.7420	0.4938	3.1913	650.4	16/38	SW:YB08_YEAST		S.PCTRPLKKMSPSPSFTSLKM.E
3.	3 / 65	2239.5761	0.4992	3.1573	432.0	14/36	GP:SPBC15C4_3		S.KDIPTKDALDTVLKYFHSF.G
4.	4 / 72	2238.7738	0.5027	3.1355	425.0	13/36	GP:SPBC16H5_3		R.QFLFRKKHVKQPKAIVAAK.T
5.	5 / 14	2238.6971	0.5334	2.9419	594.6	15/38	GPN:SPBC32F12_5		P.KDIKSIKSVPICEKHRASVV.K

```
1.  SW:YAD5_YEAST 112.3 KD PROTEIN IN PYK1-SNC1 INTERGENIC REGION [MASS=112268]

2.  SW:YB08_YEAST HYPOTHETICAL 62.7 KD PROTEIN IN RPB5-CDC28 INTERGENIC REGION
    [MASS=62686]

3.  GP:SPBC15C4_3 S.pombe chromosome II cosmid c15C4; SPBC15C4.03, len:459, SI
    MILARITY:Saccharomyces cerevisiae, YOR370C, Q06761, mrs6 protein., (594 aa
    ), fasta scores: opt: 287, E():1.5e-30, (31.2% identity in 430 aa). [MASS=
    52006]
```

Fig. 7.4. SEQUEST output-file example

If SEQUEST identifies a sequence that accurately matches the spectrum, the majority of the peaks should be aligned with predicted fragment ions. Typically, several consecutive fragment ions in a series will match observed ions. The browser interface allows visual evaluation of the match between the spectrum and the sequence.

For studies where thousands of spectra are being processed, it may not be possible to examine each match individually. The correlation scores provide a quick assessment of each match. The "XCorr" score describes how well the virtual spectrum for each sequence cross-correlates with the observed spectrum. In the example output file (Fig. 7.4), the XCorr score for the top sequence is 6.3045, indicating a very strong correlation between the observed spectrum and the virtual spectrum for this sequence. "DeltCn," however, shows the degree by which lower-ranked peptide scores differ from the correlation score of the best match. In the example output, the second-best sequence has a 0.4938 DeltCn, indicating that its correlation score is only 50.62% that of the best match. The higher the DeltCn of the second-ranked match, the more likely it is that the first match is correct.

Enzyme specificity provides an additional type of match verification. A search can be conducted without specifying which enzyme was used to cut the proteins. SEQUEST then generates its list of matching peptides without knowledge of the enzyme cleavage site. If the sequences it provides match the enzyme specificity known to the investigator (for example, a peptide match is claimed if it terminates in lysine when trypsin has been used), the credibility of the match is heightened. Matching enzyme specificity can be an independent measure of sequence-match accuracy. The scores used by SEQUEST to filter the possible database sequences can also be useful for determining the validity of a match. These scores include information about the number of ions expected and found for each sequence and spectrum. They also provide information about the contiguity of the predicted fragments; a long stretch of predicted and found peaks is a good indicator of a correct match. Because these scores are generated independently of the cross-correlation scores, the information they supply can be treated as a separate check of match accuracy.

7.3.2 Interpretation of an Entire LC/MS/MS Analysis

Information from multiple spectra can be combined to improve match accuracy. One can be more confident of a protein's presence in a mixture when multiple peptides from it are found. The SEQUEST SUMMARY program was developed to accomplish this goal by combining results from a large set of searches (Fig. 7.5). The SUMMARY program parses every search result in an LC/MS/MS run and prints out a one-line summary of each output file. This allows the user to glance through the search results without having to open and view each output file. The information provided from each output file includes the MS/MS spectrum input-file name, the mass of the top peptide identified in the search, the XCorr score, the preliminary score for the top peptide, the DeltCn, the protein reference for the top peptide and the top peptide sequence. In addition to the summary of each output file, the report also contains a consensus-analysis section that shows the protein sequences identified in a search. A consensus score is attributed to each protein; this score is primarily based on the number of times a peptide from that protein was identified in the set of output files. SUMMARY helps to consolidate all the information developed from an LC/MS/MS run into a single file.

7.3.3 Interpretation of Multiple LC/MS/MS Analyses

In some cases, the results of multiple LC/MS/MS experiments are compared. This technique can be used to highlight the differences in protein content between a control and an experimental sample. We have developed AUTO-QUEST software to automate some aspects of this type of study.

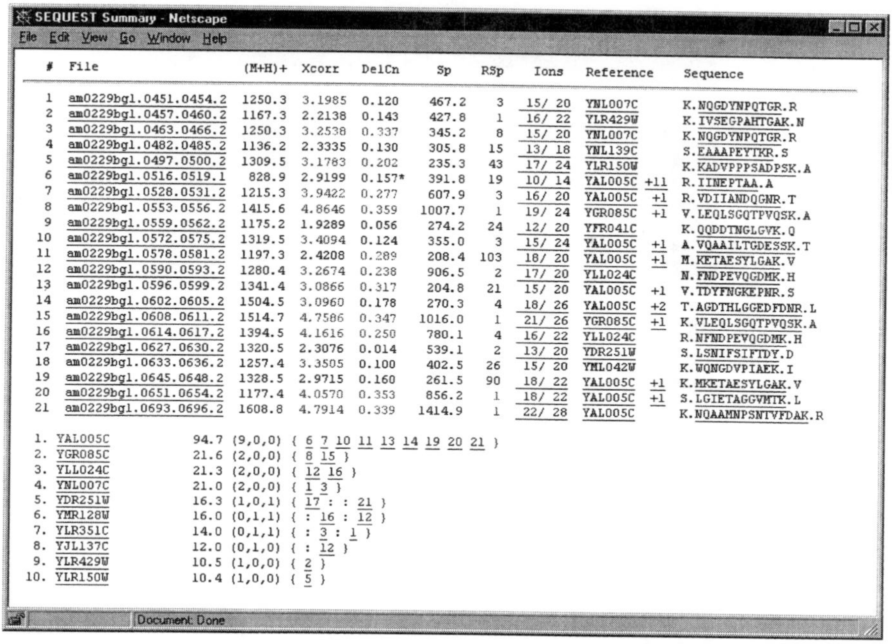

```
SEQUEST Summary - Netscape
File  Edit  View  Go  Window  Help

 #  File                  (M+H)+  Xcorr   DelCn   Sp      RSp  Ions    Reference     Sequence
 1  am0229bg1.0451.0454.2 1250.3  3.1985  0.120   467.2    3   15/ 20  YNL007C       K.NQGDYNPQTGR.R
 2  am0229bg1.0457.0460.2 1167.3  2.2138  0.143   427.8    1   16/ 22  YLR429W       K.IVSEGPAHTGAK.N
 3  am0229bg1.0463.0466.2 1250.3  3.2538  0.337   345.2    8   15/ 20  YNL007C       K.NQGDYNPQTGR.R
 4  am0229bg1.0482.0485.2 1136.2  2.3335  0.130   305.8   15   13/ 18  YNL139C       S.EAAAPEYTKR.S
 5  am0229bg1.0497.0500.2 1309.5  3.1783  0.202   235.3   43   17/ 24  YLR150W       K.KADVPPPSADPSK.A
 6  am0229bg1.0516.0519.1  828.2  9.2199  0.157*  391.8   19   10/ 14  YAL005C  +11  R.IINEPTAA.A
 7  am0229bg1.0528.0531.2 1215.3  3.9422  0.277   607.9    3   16/ 20  YAL005C  +1   R.VDIIANDQGNR.T
 8  am0229bg1.0553.0556.2 1415.6  4.8646  0.359  1007.7    1   19/ 24  YGR085C  +1   V.LEQLSGQTPVQSK.A
 9  am0229bg1.0559.0562.2 1175.2  1.9289  0.056   274.2   24   12/ 20  YFR041C       K.QQDDTNGLGVK.Q
10  am0229bg1.0572.0575.2 1319.5  3.4094  0.124   355.0    3   15/ 24  YAL005C  +1   A.VQAAILTGDESSK.T
11  am0229bg1.0578.0581.2 1197.3  2.4208  0.289   208.4  103   18/ 20  YAL005C  +1   M.KETAESYLGAK.V
12  am0229bg1.0590.0593.2 1280.4  3.2674  0.238   906.5    2   17/ 20  YLL024C       N.FNDPEVQGDMK.H
13  am0229bg1.0596.0599.2 1341.4  3.0866  0.317   204.8   21   15/ 20  YAL005C  +1   V.TDYFNGKEPNR.S
14  am0229bg1.0602.0605.2 1504.5  3.0960  0.178   270.3    4   18/ 26  YAL005C  +2   T.AGDTHLGGEDFDNR.L
15  am0229bg1.0608.0611.2 1514.7  4.7586  0.347  1016.0    1   21/ 26  YGR085C  +1   K.VLEQLSGQTPVQSK.A
16  am0229bg1.0614.0617.2 1394.5  4.1616  0.250   780.1    4   16/ 22  YLL024C       R.NFNDPEVQGDMK.H
17  am0229bg1.0627.0630.2 1320.5  2.3076  0.014   539.1    2   13/ 20  YDR251W       S.LSNIFSIFTDY.D
18  am0229bg1.0633.0636.2 1257.4  3.3505  0.100   402.5   26   15/ 20  YML042W       K.WQNGDVPIAEK.I
19  am0229bg1.0645.0648.2 1328.5  2.9715  0.160   261.5   90   18/ 22  YAL005C  +1   K.MKETAESYLGAK.V
20  am0229bg1.0651.0654.2 1177.4  4.0570  0.353   856.2    1   18/ 22  YAL005C  +1   S.LGIETAGGVMTK.L
21  am0229bg1.0693.0696.2 1608.8  4.7914  0.339  1414.9    1   22/ 28  YAL005C       K.NQAAMNPSNTVFDAK.R

 1.  YAL005C   94.7  (9,0,0)  { 6 7 10 11 13 14 19 20 21 }
 2.  YGR085C   21.6  (2,0,0)  { 8 15 }
 3.  YLL024C   21.3  (2,0,0)  { 12 16 }
 4.  YNL007C   21.0  (2,0,0)  { 1 3 }
 5.  YDR251W   16.3  (1,0,1)  { 17 : : 21 }
 6.  YMR128W   16.0  (0,1,1)  { : 16 : 12 }
 7.  YLR351C   14.0  (0,1,1)  { : 3 : 1 }
 8.  YJL137C   12.0  (0,1,0)  { : 12 }
 9.  YLR429W   10.5  (1,0,0)  { 2 }
10.  YLR150W   10.4  (1,0,0)  { 5 }

                  Document: Done
```

Fig. 7.5. SEQUEST summary output example

When a project entails the processing of thousands of spectra from multiple sources, SUMMARY's output may become lengthy. AUTOQUEST provides a web-based interface for examining multiple LC/MS/MS runs side by side (Fig. 7.6). The software groups peptides by the protein from which they came. The number of spectra matched to each peptide is reported, and the correlation score of the best matching peptide is represented by a letter code. Peptides identified from spectra of triply charged ions are marked with a plus symbol, because these spectra have unusually high correlation scores. The end of the report summarises the information, both for individual runs and multiple runs. The HTML interface allows the user to interactively include or exclude results based on the correlation score and peptide enzyme specificity. Additionally, the underlying SEQUEST output files and protein-sequence coverage information are available via hypertext links. AUTOQUEST provides an enormous amount of information at a glance. This format makes it easy to interpret the identifications made in multiple LC/MS/MS runs. To track a specific peptide, the researcher follows the row representing that peptide. The information highlights the sample in which a peptide was identified, the number of spectra matching that peptide in each run, and the significance of the best correlation for each peptide in each run. AUTOQUEST summarises peptide information from multiple experiments.

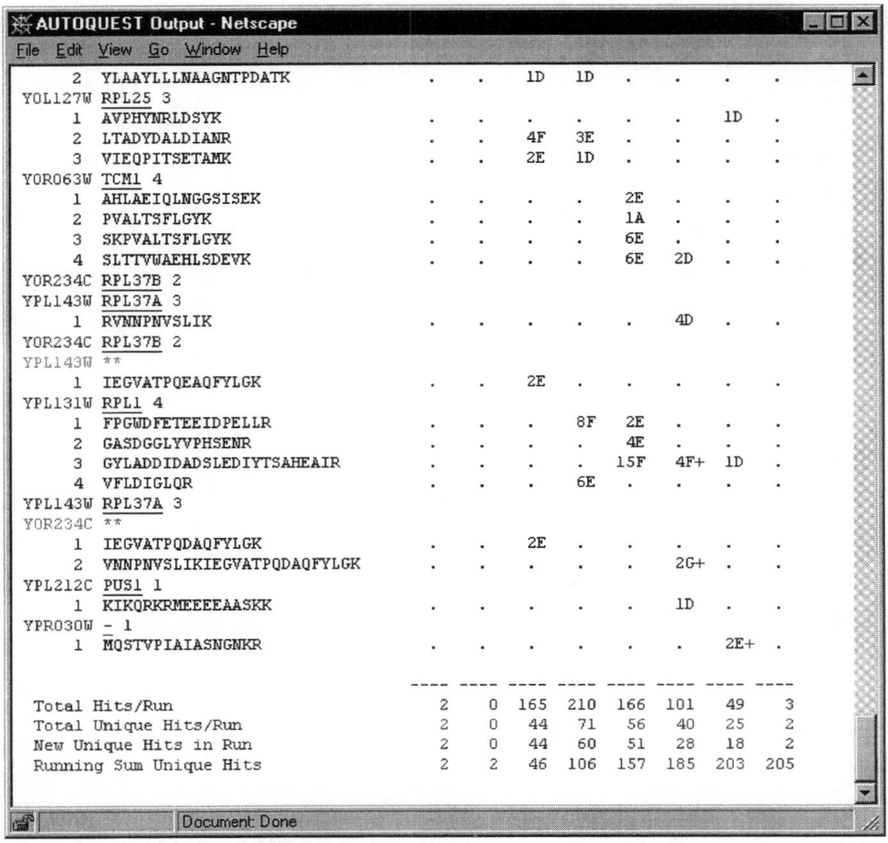

```
AUTOQUEST Output - Netscape                                    _ □ ×
File  Edit  View  Go  Window  Help

    2  YLAAYLLLNAAGNTPDATK          .    .   1D   1D    .    .    .    .
YOL127W RPL25 3
    1  AVPHYNRLDSYK                 .    .    .    .    .    .   1D    .
    2  LTADYDALDIANR                .    .   4F   3E    .    .    .    .
    3  VIEQPITSETAMK                .    .   2E   1D    .    .    .    .
YOR063W TCM1 4
    1  AHLAEIQLNGGSISEK             .    .    .    .   2E    .    .    .
    2  PVALTSFLGYK                  .    .    .    .   1A    .    .    .
    3  SKPVALTSFLGYK                .    .    .    .   6E    .    .    .
    4  SLTTVWAEHLSDEVK              .    .    .    .   6E   2D    .    .
YOR234C RPL37B 2
YPL143W RPL37A 3
    1  RVNNPNVSLIK                  .    .    .    .    .   4D    .    .
YOR234C RPL37B 2
YPL143W **
    1  IEGVATPQEAQFYLGK             .    .   2E    .    .    .    .    .
YPL131W RPL1 4
    1  FPGWDFETEEIDPELLR            .    .    .   8F   2E    .    .    .
    2  GASDGGLYVPHSENR              .    .    .    .   4E    .    .    .
    3  GYLADDIDADSLEDIYTSAHEAIR     .    .    .    .  15F  4F+  1D    .
    4  VFLDIGLQR                    .    .    .   6E    .    .    .    .
YPL143W RPL37A 3
YOR234C **
    1  IEGVATPQDAQFYLGK             .    .   2E    .    .    .    .    .
    2  VNNPNVSLIKIEGVATPQDAQFYLGK   .    .    .    .    .   2G+   .    .
YPL212C PUS1 1
    1  KIKQRKRMEEEEAASKK            .    .    .    .    .   1D    .    .
YPRO30W - 1
    1  MQSTVPIAIASNGNKR             .    .    .    .    .   2E+   .    .

                                 ---- ---- ---- ---- ---- ---- ---- ----
    Total Hits/Run               2    0  165  210  166  101   49    3
    Total Unique Hits/Run        2    0   44   71   56   40   25    2
    New Unique Hits in Run       2    0   44   60   51   28   18    2
    Running Sum Unique Hits      2    2   46  106  157  185  203  205

              Document: Done
```

Fig. 7.6. AUTOQUEST output example

Understanding the results of a SEQUEST search requires some attention at several different levels. Several scores help to clarify the relationship between an individual spectrum and a matching sequence. SUMMARY's broader look at several spectra within an LC/MS/MS run can reveal that more than one peptide from a given protein has been found. AUTOQUEST can enable the researcher to compare LC/MS/MS runs to highlight the differences between controls and experimental results. This three-level analysis is greatly aided by the software accompanying SEQUEST.

7.4 Extensions of Functionality

SEQUEST's ability to identify proteins is its best-known feature. Recent efforts, however, have focused on making it more useful for locating and identifying post-translational modifications. Ongoing research has sought to improve its performance by removing undesirable spectra and by parallelising its algorithm to span multiple computers.

7.4.1 Comprehensive Modification Analysis

SEQUEST can identify post-translational modifications of peptides. The simplest form, the "static modification," involves changing a particular residue or terminus mass in the search-parameters file. During the search, database sequences containing that residue are interpreted as having the new, altered mass. Carboxymethylation, for example, causes cysteine to gain 58 Da. Using this technique, one would expect a mass of 161 Da instead of 103 Da for cysteine in database sequences. Static modifications are useful whenever all residues of a particular type can be expected to have a mass different than usual.

A computationally expensive but more powerful form of modification analysis is termed "differential modification." To perform a differential modification search, a set of amino-acid residues and their possible modifications are specified in the search-parameters file. When multiple potential sites of modification are present in a peptide sequence, SEQUEST analyses all possible combinations of modified and unmodified residues for that peptide. This can both identify that a modification is present in a sequence and locate the specific site of a modification. An example application of a differential modification would be the search for phosphorylation, a modification associated with many biological processes, including signal transduction. A SEQUEST search for a phosphorylated peptide (including searching for phosphoserine, phosphothreonine, and phosphotyrosine) would require a differential modification search, because not all serine, threonine and tyrosine residues are phosphorylated. When both modified and unmodified forms of a residue are present, a differential-modification search is necessary.

SEQUEST itself can consider up to two sets of differential modifications and static modifications of any combination of amino acids in a single search. However, as many as 200 types of covalent modifications may exist (Krishna and Wold 1993). Automatic investigation of a larger, more comprehensive set of modifications requires a helper program. This helper program, named SEQUEST-PTM (post-translational modification), works as follows. Instead of rewriting SEQUEST to consider a large set of modifications for a single database query, SEQUEST-PTM automates the process of performing multiple database queries, with each query searching for a different individual modification. It does this by sequentially reading each modification specified in the modification list, updating the search-parameters file with the current modification, running a search on every unknown spectrum and saving the output files with unique names specific to the modification. This results in multiple SEQUEST output files for every spectrum, one for each specified modification. The multiple modification output files for each spectrum are then collated and, by comparing the results of the searches, the output file with the best peptide match is highlighted in an HTML document. This method allows multiple modifications to be searched individually, but the results of all the modification searches are analysed against each other to select the best matching modified peptide.

7.4.2 Performance Improvements

With the advent of automated, computer-controlled LC/MS/MS data acquisition, very large volumes of data can be produced by a single mass spectrometer. It is not uncommon for a single LC/MS/MS experiment to yield thousands of spectra. This count can easily rise to tens of thousands of spectra per single mass spectrometer per day through efficient use of a mass spectrometer or through use of an autosampler. In addition, sequence databases are growing at ever-higher speeds. Given the large number of spectra requiring analysis, data analysis is the limiting step in overall throughput. Individual SEQUEST searches take from a second to hours to complete, depending on many factors, including the database to be queried, the search parameters specified, and the speed of the computer processor performing the search. However, when this individual search time is multiplied by a factor of 1000 or more, one can see that analysing large data sets can easily take hours or days.

The problem of long search times is one that is created by the efficiency of mass spectrometers but can be partially addressed by various computational methods. SEQUEST has been adapted to reduce these large data sets by implementing a system to remove known contaminant spectra and low-quality spectra. To analyse the remaining spectra more quickly, SEQUEST has been modified to run on multiple computers connected in a cluster.

7.4.2.1 Contaminant Filtering

Even careful preparation sometimes cannot prevent keratin peptides from appearing as components in the sample. Whether this is a persistent problem or an occasional occurrence, analysing the spectra of contaminants can waste large amounts of user and computer time. Quickly identifying and removing contaminant spectra prior to the database search can yield substantial performance benefits.

One possible method of removing contaminant spectra would be to maintain a database of masses and charge states of commonly seen contaminant peptides. As the spectra are extracted from the mass spectrometer's capture file, spectra that match a known contaminant's precursor mass and charge-state combination could be discarded. A drawback of such a method is that it would eliminate spectra that may be informative but that coincidentally have the same precursor mass and charge states as known contaminants.

A more selective method of performing contaminant filtering is to compare spectra to a small database of contaminant sequences. Spectra matching contaminant sequences are discarded, leaving the remaining spectra to be searched against the full sequence database. Both the identification and removal of the contaminant spectra can be easily automated. However, the major drawback is that this method is sensible only if the contaminant database is much smaller than the target-sequence database to be searched. Sets of spectra including many contaminants benefit most from this technique, but using too large a database of contaminants can slow the cleanup process relative to the time gained by searching the full sequence database with fewer sequences.

A third method of contaminant filtering is to make use of a contaminant-spectrum library. This library stores MS/MS files of previously identified contaminant peptides. When SEQUEST search files are created, all spectra can be compared with files in the library. For each search file, the precursor mass of the spectrum is compared with the information stored in the contaminant library. If any spectrum contains a precursor mass that matches a stored spectral-library precursor mass, further analysis by direct comparison between the input and library spectra is performed. This direct comparison is done via spectral-correlation analysis similar to the correlation analysis performed as part of the SEQUEST search, but with the cross-correlation score normalised by the autocorrelation of the input spectrum. Any input spectra that match a library spectrum above a certain correlation threshold are discarded. As mentioned before, this method is only beneficial if the spectral-library analysis time is small. Performing many cross-correlations is computationally expensive and slow but, for small contaminant libraries, the precursor-mass filter makes correlation computations unnecessary.

7.4.2.2 Spectral-Quality Filtering

Many tandem spectra are so poor in quality that they cannot be matched to database sequences. These spectra may result from unusual peptide fragmentation or from the presence of too little of a particular peptide in a sample. By removing these spectra from the pool to be analysed by SEQUEST, a search can be accelerated similar to the way in which contaminant filtering speeds the search. Contaminant filtering removes spectra that match known contaminant spectra. Removing poor spectra, however, targets those that do not match known spectra through a very different process. Quality filtering can be more beneficial to SEQUEST's speed than contaminant filtering because of the large number of poor spectra in a typical LC/MS/MS run.

The quality-filtering process can encompass a variety of rules. One simple rule sets a minimum threshold for a spectrum's summed ion-current intensity and/or maximum peak intensity. This eliminates spectra resulting from peptide ions that are too low in abundance to produce meaningful fragmentation spectra. Another filtering rule requires a minimum and/or maximum number of peaks in a spectrum. A poor spectrum, for example, could have only three or four high-intensity peaks that would pass the first rule but which would fail the second rule. Similarly, spectra that contain too many peaks often result from noise and fail to produce reliable identifications. Other possible rules could consider relative peak intensities and the location or distribution of peaks in a spectrum. The application of such rules can rapidly remove many spectra from consideration that would otherwise slow SEQUEST.

These rules should be very conservative, because SEQUEST has the ability to generate positive identifications with poor spectra. Nevertheless, the use of even simple filtering rules will remove a significant number of MS/MS spectra from an LC/MS/MS experiment. The number of spectra removed can be as high as 40% or more, significantly reducing the data-analysis time required.

7.4.2.3 High-Performance Computing Using a Parallel Cluster

As data-acquisition throughput and database sizes continue to grow, so does the need for powerful data analysis hardware. Even the fastest single-processor workstations may not provide the performance required for high-throughput laboratories. For these situations, the use of more than one processor computing in parallel is a logical solution. Computers in which multiple processors share the same physical memory are available from many vendors. However, shared-memory multiprocessor (SMP) machines with more than four processors are extremely expensive. Although many large organisations have access to supercomputer-class SMP boxes, the prohibitive cost of such hardware puts them out of the reach of many individual users and small laboratories.

An increasingly popular alternative to SMP machines was pioneered by the Beowulf Project, featuring a dedicated cluster of commodity computers connected by a high-performance network (Sterling et al. 1995). Memory is local for each node, and communication occurs through the network using messages. There is usually one master node that acts as the user console and is the cluster's gateway to the outside world. Since there is little (if any) direct user interaction with the other nodes in the cluster, the master node is typically the only one with a dedicated monitor, keyboard and mouse. The advantage of using a cluster to perform multiprocessing is that the cost is usually many times lower than equivalently performing SMP machines. Making proper use of such a cluster typically requires some modifications of a program and is only useful for some types of applications. Fortunately, database searching is a type of problem that is well suited for this type of arrangement.

SEQUEST is usually run on several thousand spectra. Because the algorithm repeats the same process for each spectrum, SEQUEST is amenable to alteration for use with clusters. The master node starts the process by sending parameters that characterise the type of search to be performed. Each computer in the cluster is given a spectrum on which to run SEQUEST. When a computer finishes processing its spectrum, it notifies the master node, which then sends the computer another spectrum to be processed. This continues until the entire set of spectra has been analysed. The entire process of concurrent processing is transparent to the user.

The cluster that we assembled to perform SEQUEST searches is made up of a total of 12 nodes. Each node contains a Compaq Alpha 21164A 533-MHz processor, 64 MB of random-access memory (RAM), an IDE hard drive and a 100 Mb/s Ethernet card with a DEC Tulip chip set. The master node has an additional 256 MB of RAM and second Ethernet card. The operating system installed on each node is Red Hat Linux, and message passing is accomplished using the Parallel Virtual Machine software (Geist et al. 1994). The Alpha processor was selected because of a combination of the high computing performance it offers and its availability through a number of third party vendors. To maximise performance, local copies of search databases are stored on each node. Example run times for the cluster researching 500 MS/MS spectra in two sets of databases using 1, 6 and 12 of the central processing units (CPUs) are shown in Table 7.1.

The searches were performed with a 1.0-amu peptide mass tolerance, and no enzyme was specified. As can be seen, the search times scale linearly with the number of processors utilised in the cluster. Effective throughput of the cluster is less than 13 s per full non-redundant protein-database search (352,925 sequences) and was less than 0.4 s for the yeast database (6351 sequences). Increasing the performance of the cluster is as simple as adding additional computers. Since the parallel implementation of SEQUEST allows the nodes to run asynchronously, the system will run optimally (with every

Table 7.1. SEQUEST Alpha-computer cluster search times (500 spectra; peptide mass ± 1.0 amu)

	1 CPU (hh:mm:ss)	6 CPUs (hh:mm:ss)	12 CPUs (hh:mm:ss)
Yeast ORFs (6351 proteins)	00:37:09	00:05:34	00:02:53
Non-redundant (352,925 proteins)	10:35:57	03:16:38	01:47:21

CPU, central processing unit; ORF, open reading frame

CPU continuously processing at 100% utilisation) whether the new CPUs are faster or slower than the existing nodes.

SEQUEST's versatility and performance are continually being improved. Because of its ability to identify major protein modifications, SEQUEST can be an enormously useful tool for identifying sites of post-translational control in proteins. As research in contaminant and low-quality spectrum filtering continues, the algorithm will be better focused on spectra of interest, and its speed can be multiplied through the use of clustered computers. SEQUEST's functionality can be extended in power and performance.

7.5 Applications of SEQUEST

MS/MS has been adapted to accept peptides from many sources. On-line multidimensional chromatography can provide powerful separations of peptides before they reach the mass spectrometer. Techniques allowing the extraction of peptides from standard electrophoretic gels allow many biochemical procedures to be used for preliminary purification. SEQUEST can be used to identify the peptides produced by a wide array of techniques.

7.5.1 Complex-Mixture Analysis

SEQUEST can be used to catalogue the peptide components of complex protein mixtures. We used this approach to identify the proteins enriched from the periplasmic space of *Escherichia coli* (Link et al. 1997). Proteins localised to the periplasmic space were enriched by the method of Neu and Hoppel (1965). The collection of proteins was partially fractionated by ion-exchange chromatography. The fractions were digested with trypsin and then analysed by LC/MS/MS. SEQUEST identified 80 proteins from the mixture. Less than half of the proteins contained the periplasmic signal-peptide sequence, and some of the other proteins were abundant cytoplasmic proteins that could have leaked from cells ruptured during the enrichment process. A number of the remaining proteins were candidates for localisation in the periplasmic space. Use of SEQUEST enabled us to rapidly and efficiently identify a large number of proteins from a complex mixture.

7.5.2 Subtractive Analysis

SEQUEST can be used as a subtractive analysis tool to identify peptide spectra for proteins not in the database. We used this technique to identify haemoglobin variants. Peptide spectra for mutant forms of the protein were isolated from a large data set by removing all spectra matching sequences in a database. After this isolation, the remaining spectra were analysed by the

SEQUEST-SNP program to identify peptides that varied by single residues from known sequences. The analysis found three variants. In this way, spectral comparison can be used to remove known sequences, yielding a smaller set of data to enable unknown or modified peptide spectra to be analysed by SEQUEST.

7.5.3 Case Studies from in-Gel Digest Analysis

SEQUEST can be used to identify proteins purified by gel separation. Mass spectrometers are sensitive enough to produce high-quality tandem spectra from gel spots detectable only through silver staining. SEQUEST has identified proteins purified by activity assay and by their interactions with known proteins.

Monitoring a specific activity throughout an enrichment process is a traditional biochemical approach that was used to identify a regulatory factor of insulin-like growth factor (IGF; Lawrence et al. 1999). IGFs are proteins with anabolic and mitogenic activities and are important growth-regulating proteins. One potent inhibitor of IGF stimulatory effects is IGF BP-4. The binding activity and inhibitory effect of IGF BP-4 can be reversed through proteolysis. To identify the protease responsible, IGF BP-4 specific proteolytic activity was followed throughout a purification, which yielded three sodium dodecyl sulphate polyacrylamide-gel electrophoresis (SDS-PAGE) bands. SEQUEST identification of the proteins in one band revealed a pregnancy-associated plasma protein that had not been known to possess this activity. Further functional characterisation of the protein identified it as the IGF BP-4 regulatory protease. SEQUEST can provide unambiguous identification of proteins in studies tracing specific activities.

Knowledge of one protein involved in a process can be used to produce information about interacting proteins. McCormack et al. (1997) illustrated that MS can be used for identification of proteins obtained by immunoprecipitation, protein affinity-interaction chromatography and proteins enriched through interaction with a core protein complex. Investigation of Nijmegen breakage syndrome made use of this capability (Carney et al. 1998). The disease can yield increased incidence of blood-borne leukaemias, immunodeficiencies and other physical abnormalities associated with DNA breakage. Carney et al. precipitated proteins known to repair DNA double-strand breaks and found that several other proteins accompanied the known proteins. After separation by SDS-PAGE, a protein of 95 kDa was removed from the gel and digested for analysis by LC/MS/MS. The collection of tandem mass spectra was used to search sequence databases, including the expressed-sequence tag (EST) database. A match to an EST sequence led to the identification of a particular gene at a locus known to be associated with the syndrome. Further experiments verified the gene's role in DNA-breakage repair. Identifications are possible even when the genome of interest is not yet completely known. SEQUEST provides a powerful analytical tool for many types of protein studies.

7.6 Conclusion

SEQUEST enormously simplifies the task of peptide MS/MS. By providing automated analysis of tandem mass spectra, the software enables the researcher to dramatically increase the scope and throughput of mass-spectrometric investigations. Groups who have previously avoided mass spectrometers because of their complexity can get genuine benefits from their use without the difficult learning curve. SEQUEST has much to offer to the biologist studying proteins and their interactions.

By providing a link between sequence databases and observed spectra, SEQUEST provides the same benefits to protein research that automated sequencing brought to DNA research. The identification of proteins can be done far more rapidly than before, and the certainty of those identifications can be quantified. By removing this barrier to protein research, SEQUEST provides a valuable tool for practical proteomics.

References

Carney JP, Maser RS, Olivares H, Davis EM, Le Beau M, Yates JR III, Hays L, Morgan WF, Petrini JH (1998) The hMre11/hRad50 protein complex and Nijmegen breakage syndrome: linkage of double-strand break repair to the cellular DNA damage response. Cell 93:477–486

Eng JK, McCormack AL, Yates JR III (1994) An approach to correlate tandem mass spectral data of peptides with amino acid sequences in a protein database. J Am Soc Mass Spectrom 5:976–989

Geist A, Beguelin A, Dongarra J, Jiang W, Manchek R, Sunderam V (1994) PVM: Parallel Virtual Machine a users' guide and tutorial for networked parallel computing. MIT Press, Cambridge

Krishna RG, Wold F (1993) Post-translational modification of proteins. Adv Enzymol Relat Areas Mol Biol 65:265–298

Lawrence JB, Oxvig C, Overgaard MT, Sottrup-Jensen L, Gleich GJ, Hays LG, Yates JR III, Conover CA (1999) The insulin-like growth factor (IGF)-dependent IGF binding protein-4 protease secreted by human fibroblasts is pregnancy-associated plasma protein-A. Proc Natl Acad Sci USA 96:3149–3153

Link AJ, Carmack E, Yates JR III (1997) A strategy for the identification of proteins localized to subcellular spaces: application to E. coli periplasmic proteins. Int J Mass Spectr Ion Proc 160:303–316

McCormack AL, Schieltz DM, Goode B, Yang S, Barnes G, Drubin D, Yates JR III (1997) Direct analysis and identification of proteins in mixtures by LC/MS/MS and database searching at the low-femtomole level. Anal Chem 69:767–776

Neu HC, Heppel LA (1965) The release of enzymes from Escherichia coli by osmotic shock and during the formation of spheroplast. J Biol Chem 240:3685–3692

Powell LA, Heiftje GM (1978) Computer identification of infrared spectra by correlation-based file searching. Anal Chim Act 100:313–327

Sterling T, Becker DJ, Savarese D, Dorband JE, Ranawak UA, Packer CV (1995) Beowulf: a parallel workstation for scientific computation. Proceedings of the International Conference on Parallel Processing, p 95

8 Interpreting Peptide Tandem Mass-Spectrometry Fragmentation Spectra

WERNER STAUDENMANN and PETER JAMES

8.1 Introduction to Peptide Tandem Mass Spectrometry

8.1.1 Peptide Ionisation

Mass spectrometers separate molecules according to their mass-to-charge ratio (m/z) using magnetic fields to deflect the ions [magnetic-sector mass spectrometry (MS)], crossed voltage (direct current) and radio-frequency (RF) emissions to create stable trajectories (quadrupole or ion-trap MS) or mass-selective acceleration to separate the ions in space [time-of-flight (TOF) MS]. All of these techniques require the ions to be in a vacuum (between 10^{-3} and 10^{-8} Torr). Thus, the main problem is moving the molecules to be analysed into the gas phase in an ionised state.

The mass spectrometry of peptides became popular with the introduction of fast atom bombardment (Barber et al. 1981) during the early 1980s. This technique allowed the ionisation of many non-volatile biological compounds as intact molecules (up to a mass of approximately 8000 amu) without derivatisation. The peptide sample is mixed with a liquid matrix (such as glycerol) under acidic conditions and is exposed to a beam of 3-kV neutral gas atoms (often xenon) or ions (cesium). The protonated peptides are ejected (sputtered) into the gas phase and are accelerated away from the target along a voltage gradient. The initial kinetic energy imparted by the bombardment is approximately 10–30 eV (1 eV = 1.6×10^{-19} J), enough to cause some fragmentation of the peptides (the energy of a peptide bond is approximately 1 eV). By increasing the energy of the bombarding projectiles, sequence-specific fragmentation was observed (Barber et al. 1981). However, for sequencing purposes, this method requires that the peptide be pure and suffers from interference from a background of matrix ions.

Two other ionisation methods have since become standard: matrix-assisted laser desorption and ionisation (MALDI; Karas and Hillenkamp 1988) and electrospray ionisation (ESI; Fenn et al. 1989), and its close relative, atmospheric-pressure ionisation. The initial kinetic energy imparted to the ionising peptides approximately 0.1 eV for ESI and 1 eV for MALDI, so virtually no fragmentation is observed. During laser ionisation, however, the initial energy is low, but the ions are accelerated into the expanding plume of matrix, which can cause collisional fragmentation. If the instrument is being operated in linear mode, there is no separation of daughter ions from parent ions, and fragmentation is not easily observed (Chap. 3).

8.1.2 Peptide Fragmentation

The development of tandem mass spectrometry (MS/MS) – the coupling of two mass-selective devices, most commonly magnetic-sector (Gross 1990) or quadrupole instruments (Yost and Boyd 1990) – greatly expanded the usefulness of the technique. The first MS device serves to isolate the target peptide from other peptides and matrix ions before the peptide (the parent ion) is accelerated through a region (the collision cell) where it undergoes collisions with a gas (such as argon). The resulting fragment ions (the daughter ions) are analysed in the second MS device. Two fragmentation-energy regimes (high and low) that differ in the amount of energy used for the fragmentation, the types of ions produced and the sequence information available can be distinguished. The various types of ions that are produced by low-energy collisionally activated (or induced) dissociation (CAD or CID) are shown in Fig. 8.1. The nomenclatures shown are those proposed by Roepsdorrf and Fohlmann (1984) and modified by Biemann (1990). The magnetic-sector and TOF instruments use high voltages (>10 kV) to accelerate the ions giving "high-energy" collisions of 5000–30,000 eV, which produces peptide-backbone and side-chain cleavage.

The amount of energy transferred in a collision is very small, approximately 1–15%. The quadrupole and ion-trap instruments produce "low-energy" collisions of approximately 10–30 eV (much less than the 1000 eV required for side-chain cleavage). Although the energy transferred is only approximately 0.1 eV, peptide-bond cleavage occurs, because the gas pressure in the collision cell is set so that the ions undergo multiple collisions and can accumulate the necessary energy as they travel slowly through the collision chamber. The effectiveness of fragmentation depends on the number of collisions (gas pressure), the energy deposition (the greater the atomic cross-sectional area of the gas, the larger the transfer; i.e. Xe > Ar > He) and the charge on the ion. In a triple-quadrupole or quadrupole-TOF (Q-TOF) mass spectrometer, the collision energy is set according to the charge and mass of a ion. For example, approximately 50 eV gives the same collision energy to a singly charged ion of mass 2000 amu that 25 eV imparts to a doubly charged ion with a mass of 2000 amu (given a constant gas pressure, usually 2.5 mTorr argon). It is possible to fragment peptides by allowing them to collide with a metal surface (Mabud et al. 1985), which results in a very large deposition of energy (7 eV for 100-eV collisions, even in a triple-quadrupole or ion-trap mass spectrometer). The ion-trap mass spectrometer (in contrast to the triple-quadrupole mass spectrometer) operates under a constant pressure (1–2 mTorr) of helium, and collisional activation is brought about by causing the ion to vibrate at its resonant frequency, causing it to undergo multiple low-energy collisions.

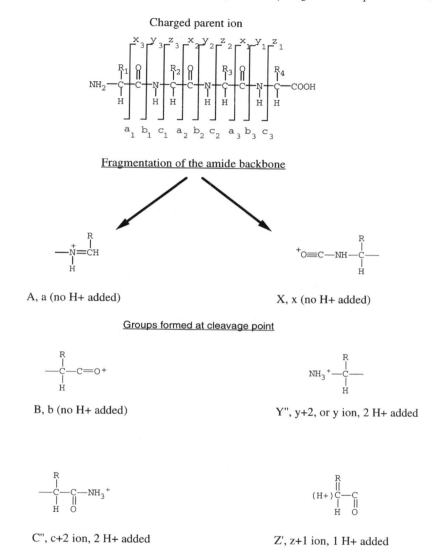

Fig. 8.1. Ion-series nomenclature

We will restrict our discussion of fragmentation to low-energy regimes, since this is the mode used by most instruments in protein-chemistry laboratories. Readers interested in high-energy fragmentation regimes should refer to the excellent review by Papayannopoulos (1995). The advantage of high-energy collisions using currently available instrumentation is questionable and possibly will become redundant given the vast DNA databases that are being created. The ability to distinguish between isobaric amino acids, such as Leu and Ile (Johnson et al. 1987), is not absolute, because peptides containing more than one of these can give equivocal results. High-resolution instruments, such as Q-TOF hybrids, can distinguish between Lys and Gln without resorting to chemical modification or high-energy regimes (Morris et al. 1996).

8.1.3 Mechanism of Amide-Bond Cleavage

In the liquid or solid phases used for ESI and MALDI, the peptides exist as cations with positive charges (in the forms of protons localised on the N-terminus and basic side chains, such as Arg, Lys and His). The acidic residues (Glu and Asp) are protonated and, hence, are neutral. Only phosphate groups maintain a negative charge if the pH is not less than 1.5. Solvation delocalises the charge and stabilises the charged species. In the gas phase, in the absence of solvent molecules, charge transfer occurs by direct contact between carriers. Since the gas-phase proton affinity of the peptide-backbone amide linkage is equivalent to that of a primary amine, chain folding serves to allow the free movement of protons from the N-terminus and Lys side chains, resulting in an even distribution along the backbone.

When the peptide ions selected by the first mass spectrometer reach the collision cell, they undergo multiple (one to ten) collisions with the argon atoms, and the kinetic energy is converted into vibrational energy. The energy is then sufficient to cleave one of the amide-backbone bonds. The subsequent fragmentation pattern depends on the collisional energy, the pressure and type of the collision gas, the number of charges being carried by the peptide, and the amino acid sequence of the peptide (Hunt et al. 1986). If the collision energy is relatively high, then fragmentation involving a single bond cleavage is favoured, and b-type ions are preferentially formed. However, if the energy is relatively small, a pathway with a lower activation energy involving a simultaneous bond breakage and formation is favoured, and y-type ions are preferentially formed (Fig. 8.2). If the peptide is singly charged, then one of the daughter products formed is charged (an ion); the other is a neutral molecule. The neutral molecule does not respond to either RF or magnetic fields and is lost.

Therefore, the distribution of charge plays an essential role in determining the relative amounts of b- and y-type ions that are detected; it is also dependant on the number and position of Pro, His, Trp, Arg and Lys residues, which have a high basicity in the gas phase. In the low-energy regime in a triple-quadrupole mass spectrometer, the neutral and charged products do not separate immediately after bond cleavage; they remain in close contact due to dipole-charge attractions until further collisions cause separation. During this time, proton transfer to the residue with the highest basicity can occur, hence

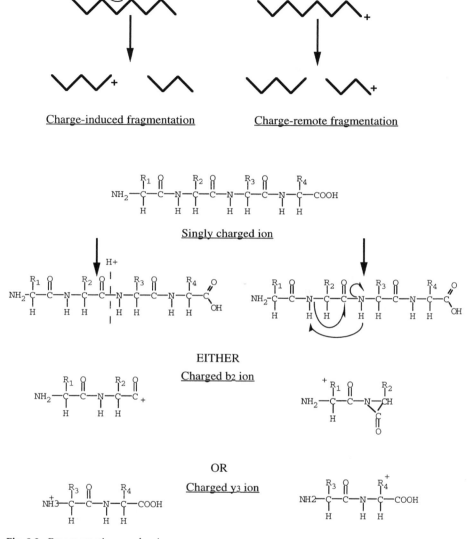

Charge-induced fragmentation   Charge-remote fragmentation

Singly charged ion

EITHER
Charged b2 ion

OR
Charged y3 ion

Fig. 8.2. Fragmentation mechanisms

explaining the relative dominance of y-type ions in tryptic peptides that usually have Arg or Lys as the C-terminal residue. This is not the case for the products in an ion trap because, if the daughter ions remain as a pair, they will still be excited and undergo rapid separation. Hence, there are proportionately more b-type ions found in a spectrum of a peptide obtained using an ion-trap mass spectrometer than with the same peptide in a triple-quadrupole mass spectrometer.

8.1.4 Sequence-Dependant Fragmentation Types

8.1.4.1 Immonium Ions

The low-mass region (m/z < 200) of MS/MS spectra often contains ions that are indicative of the presence of specific amino acids in the peptide. These immonium ions arise from at least two internal bond cleavages and are labelled with the single-letter codes for the parent amino acid (Fig. 8.3A). Table 8.1 gives a list of immonium-ion masses and other associated low-mass ions (the most important for interpretation are those due to His and Pro, which are fortunately usually quite intense).

8.1.4.2 Internal Cleavage

A second set of ions that are formed by the cleavage of two internal bonds are those that result from multi-point cleavage of the backbone. These have the same formula as b-type ions and are indicated by $(b_1y_3)_2$ (a b-type ion arising from cleavage at position 1 from the N-terminus and a y-type ion from

Table 8.1. Immonium and related ions

Amino acid	Immonium ion (m/z)	Related ions (m/z)	Comments
Ala	44		Marginally useful
Arg	129	70, 87, 100	m/z = 129; others weak (4:2:1)
Asp	88		Often weak or absent
Asn	87		Often weak
Cys	76		Relatively weak
Gly	30		Not useful
Gln	101		m/z = 101; often weak
Glu	102		Very strong
His	110	81	Strong 110:81 (3:1)
Ile/Leu	86	84	Strong 86:84 (4:1)
Lys	101		Usually weak or absent
Met	104	61	Strong 104:61 (3:4)
Phe	120	91	Strong 120:91 (3:1)
Pro	70		
Ser	60		
Thr	74		
Trp	159	130	Strong 130:159 (1:2)
Tyr	136	107	Strong 136:107 (1:2)
Val	72		Fairly strong

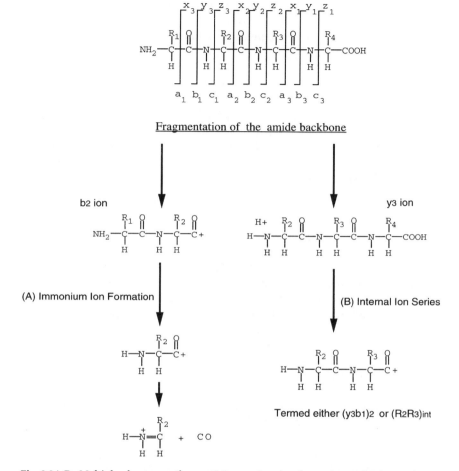

Fig. 8.3A,B. Multiple-cleavage pathways. **A** Immonium-ion formation and **B** internal-ion series

cleavage at position 3 from the C-terminus, containing two residues) or, more simply, by the amino acid sequence (Fig. 8.3B). These cleavages rarely occur, except when Pro or His (and Trp) residues are present and then they dominate the spectrum.

8.1.4.3 Preferential Cleavage C-Terminal to Asp

The carboxylic acid group of Asp is often involved in rearrangement reactions (such as the formation of β-aspartate or cleavage at Asp–Gly bonds in solution) and, in the gas phase, the group can attack the amide bond, giving rise to a cyclic intermediate that decomposes to yield a b-type ion that often dominates the MS/MS spectrum (Fig. 8.4a). The opposite situation is found for Pro, where virtually no C-terminal cleavage is found.

8.1.4.4 Low-Energy Production of a-Type Ions

Most of the a-ions observed in low energy MS/MS spectra are probably not formed by direct cleavage of the peptide bond (as observed in high energy spectra), and the corresponding x-type ions are almost never seen (Fig. 8.4b). They are formed by the loss of ammonia, causing the formation of a- and b-type ion doublets separated by 28 amu. These can be helpful when assigning ions to families during interpretation.

8.1.4.5 Loss of C-Terminal Amino Acids

Very often, when the C-terminal amino acid residue is hydrophobic (most commonly Phe and Tyr and, to a lesser extent, Leu, Ile and Val), an ion corresponding to the loss of the residue mass is observed in addition to the y-type ion. This is due to an intramolecular rearrangement (Fig. 8.4c). All these observations were described by Hunt et al. (1986).

8.1.4.6 Amino Acid-Specific Mass Losses

High-energy collisions often result in the cleavage of the bond between the β- and γ-carbon atoms of the amino acid side chain at which backbone cleavage has occurred, giving rise to d-type ions (if N-terminal) and w-type ions (if C-terminal); these are labelled a or b, depending on which part of the side chain is cleaved. Occasionally, the entire side chain is lost, producing v-type ions, though this only occurs with C-terminal fragments (Johnson et al. 1988). Low-energy collision regimes do not produce these ions; instead, the spectra often exhibit predominant ions termed b^0 or y^0 (formed by the loss of water, especially if the N-terminus is Glu) and b^* and y^* (if ammonia is lost, especially if the N-terminus is Gln). These rarely occur in high-energy spectra.

The loss of water often occurs from Ser and Thr residues, and the appearance of a ladder of doublets separated by 18 amu is diagnostic. Tyr does not exhibit the loss of water to any appreciable extent. However, the loss of 18 amu is also seen when glutamic acid is at the N-terminus as a result of pyroglutamate formation. In a similar manner, Cys exhibits the loss of 34 amu due to H_2S. Very often, Cys is deliberately chemically modified to increase the digestion efficiency and prevent disulphide-bond formation. The two most common modifications, with vinylpyridine to form pyridylethylcysteine or with

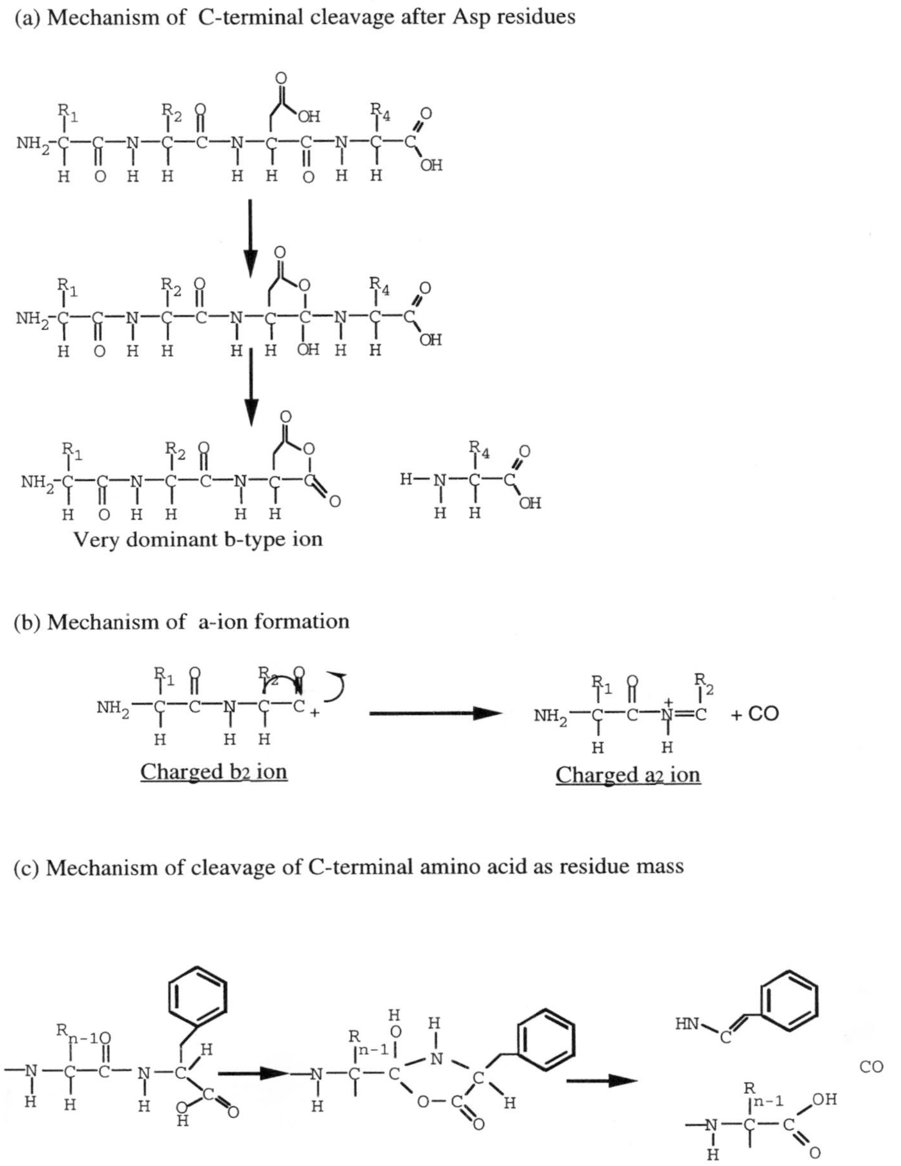

(a) Mechanism of C-terminal cleavage after Asp residues

Very dominant b-type ion

(b) Mechanism of a-ion formation

Charged b_2 ion

Charged a_2 ion

(c) Mechanism of cleavage of C-terminal amino acid as residue mass

Fig. 8.4a–c. Site specific cleavages. **a** Cleavage after Asp residues. **b** A-ion formation in a low-collision energy regime and **c** cleavage of the C-terminal amino acid as a residue mass

iodoacetic acid (or iodoacetamide) to give carboxymethylcysteine, show losses of 105 and 92 amu, respectively. Alternatively, the Cys can easily be oxidised to cysteic acid using a mixture of hydrogen peroxide in formic acid, though this can cause confusion by oxidising Met, which then has the same residue mass as Phe. Another common loss seen is 48 amu from Met-containing peptides; this is the loss of methylsulphide. The specific loss of certain groups can be used to selectively detect peptides containing certain modifications, such as phosphorylation or glycosylation, and is dealt with in more detail in Chapters 10 and 11.

8.2 Manual Interpretation of Peptide MS/MS Spectra

8.2.1 General Considerations

The ease of interpretation of an MS/MS spectrum depends, to a certain degree, on which type of instrument is used to accumulate the data. In the following examples, we will concentrate on the most common type of instruments: triple-quadrupole and ion-trap mass spectrometers with an ESI. Currently, however, the easiest spectra to interpret are produced by the (ESI) Q-TOF instruments, because they are capable of producing very high-resolution and high-mass-accuracy full-range daughter-ion spectra. One can, for example, easily distinguish the charge states of all the product ions and deconvolve the spectrum to one containing only singly charged ions, making the interpretation of MS/MS spectra of triply and quadruply charged ions possible. Moreover, the mass accuracy is sufficient to allow one to differentiate between Gln and Lys solely on the basis of the mass difference (Q = 128.06 and K = 128.09). Since the device is non-scanning, a much higher number of ions are measured during a given time period compared with the number measured by a beam instrument like a triple quadrupole mass spectrometer, thus greatly increasing the sensitivity (Morris et al. 1996).

The ion-trap mass spectrometer is also non-scanning but is limited in the number of ions it can trap between the quadrupoles without losing mass accuracy and resolution because the ions are so dense that they affect each other (space charging). The other disadvantage of the trap is its limited mass range for daughter-ion analysis (McLuckey et al. 1994). Currently, a daughter-ion range as low as 25–30% of the parent-ion mass can be analysed. In practical terms, however, because the trap is usually interfaced with an ESI source, ions of mass 1000 amu would be analysed as a doubly charged ion at an m/z of 500, giving an effective lower range of 140 amu. This drawback might soon be overcome by using pulsed heavy gases, such as argon or xenon, as the collision gas, which would have the dual effect of (1) increasing the MS/MS efficiency due to a larger collisional cross-section (and, hence, energy transfer) in comparison to helium and (2) decreasing the lower mass limit of daughter-ion retention so that it includes all immonium ions (Vachet and Glish 1998). The greatest advantage of the trap lies in its ability to retain the daughter ions after MS/MS so that a daughter ion can be selected for further MS/MS analysis (MS^n). This is dealt with further in Section 8.3, as a method to automate the collection of a complete ion series for peptides. The ion trap is usually operated using a parent-ion window of 4 amu (i.e. parent mass ±2 amu) and a daughter resolution capable of maintaining isotopic resolution (allowing one to differentiate between singly and doubly charged ions but not doubly and triply charged ions). The triple-quadrupole mass spectrometer is usually operated in low-resolution mode, extending the parent-ion selection window to give a width of 5–6 amu at half height (daughter ions 1 and 2), which greatly increases the ion transmission and, hence, the sensitivity.

8.2.2 An Approach to Manual Interpretation

When starting manual interpretation of any type of spectrum, one should round all the mono-isotopic peak masses to the nominal mass (the mono-isotopic mass rounded to the next lowest integer value). To ease calculations, the residue masses used are also nominal (Table 8.2 for accurate masses). In order to explain the general strategy developed by Don Hunt and used by the authors, we will refer to the simple MS/MS spectrum of a small tryptic peptide shown in Fig. 8.5. The accurate mass of the peptide was first determined at high resolution before and after acetylation. The native MS/MS (Fig. 8.5a) and methylated MS/MS spectra (Fig. 8.5b) were then acquired. The methylation data gives the number and positions of acidic groups in the peptide (if the derivatisation is too harsh, Gln and Asn may also become methylated). Using the mass shifts, the C-terminal carboxyl group containing y-type ions can be identified by a mass shift of 14 amu (15–1, the proton). It also helps to distinguish between Asp and Asn and Glu and Gln under low-resolution conditions. The acetylation data gives the number of Lys residues (mass shift: 42 = 43 − 1) and helps to distinguish between Lys and Gln and to determine if the N-terminus is free.

8.2.2.1 Interpretation Checklist

1. Accurately measure the masses of native parent, methylated and acetylated ions.
2. Depending on parent charge, deconvolve the spectrum to all singly charged ions. Remove all spikes.
3. Print the spectrum as is and with all regions scaled to the same size.

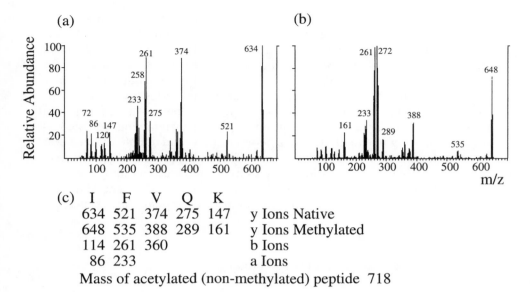

(c)
```
      I    F    V    Q    K
     634  521  374  275  147    y Ions Native
     648  535  388  289  161    y Ions Methylated
     114  261  360              b Ions
      86  233                   a Ions
```
Mass of acetylated (non-methylated) peptide 718

Fig. 8.5a–c. Interpretation of tryptic peptide tandem mass spectra. A triple-quadrupole tandem mass spectrum of a tryptic peptide in the **a** native and **b** methylated states. **c** The sequence, together with the most important sequence ions

4. Compare the native and methylated spectra and label methylation states of fragments (+1, +2. . .).
5. Label possible a/b pairs (a = b − 28).
6. Label −18 and −17 losses of water and ammonia, respectively.
7. Examine the low-mass end for R/K y_1 ions (175 or 147 amu) if tryptic.
8. Look for Pro, His and Trp immonium ions (Table 8.1).
9. If Pro, His and Trp is present, look for an intense internal-cleavage series.
10. Inspect the high-mass end for b- and y-type ions and residue-mass loss (Sect. 1.4.5).
    – The last b_{ion} has a mass of MH^+–18–residue mass.
      The first y_{ion} has a mass of MH^+–residue mass.
    – Ignore the ions between MH^+ and MH^+–60 (these represent the loss of water, ammonia, HCOOH and CH_3COOH).
11. Look for ion series by sequentially subtracting residue masses (Table 8.2).
12. Look for corresponding b/y-type ion pairs (mass of y + b = parent mass + 1).
13. For each residue, look for diagnostic losses and immonium ions.
14. Be careful of dipeptide masses that match residue masses (Table 8.3).
15. Assign a tentative sequence, taking the acetylation data into account.
16. Compare the theoretical masses to the data. Are all ions accounted for? If not, repeat the process.

Table 8.2. Residue masses

Residue	Mass (amu)	Exchangeable hydrogens	Possible equivalent mass
Gly	57.02	1	
Ala	71.04	1	
Ser	87.03	2	
Pro	97.05	0	
Val	99.07	1	AcGly
Thr	101.05	2	
Cys	103.01	2	
Pyro-Glu	111.03	1	
Leu/Ile	113.08	1	AcAla
Asn	114.04	3	Gly–Gly
Asp	115.03	2	
Lys	128.09	3	Gly–Ala
Gln	128.06	3	Gly–Ala
Glu	129.04	2	AcSer
Met	131.04	1	
His	137.06	2	
Phe	147.07	1	
Arg	156.10	5	Gly–Val or AcAsn
CmCys	161.01	2	
Tyr	163.06	2	
Trp	186.08	2	Gly–Glu, Ala–Asp or Ser–Val

8.2.3 Interpreting an MS/MS Spectrum of a Simple Tryptic Peptide

We will now examine Fig. 8.5, applying the series of steps listed in Section 8.2.2. The masses of the native, methylated and acetylated peptides are given in Fig. 8.5 (634, 648 and 718 amu, respectively). Since these are all singly charged ions, there is no need to deconvolve the spectrum. Comparing the native and methylated spectra, we see that virtually all of the ions increase their mass by 14 amu. Since this is a tryptic peptide, there cannot be any acidic residues in the sequence since only one methyl group is being added. Therefore, all the ions labelled must be in the y series. The only ions not increasing by 14 amu are the immonium ions and those with masses of 114, 261 and 360 amu, which all have an accompanying ion 28 amu lower and, hence, are probably b- and a-type ions. The only strong loss of 18 or 17 amu is seen with 275-amu native/289-amu methylated and 374-amu native/388-amu methylated ions, indicating the possible presence of K, N, Q or R. There is an ion at 147 amu and none at 175 amu, so the C-terminal residue is probably K (which agrees with what one would expect when cleaving with trypsin). There are no intense immonium ions that would indicate internal fragmentation (Pro or His). Inspection of the high-mass end shows that there is indeed a weak b-type ion corresponding to b_4 Lys (mass: $634 - 128 - 18 = 488$ amu). However, the first major ion is 521 amu, which

Table 8.3. Dipeptide masses. All masses are measured in Daltons

	Gly	Ala	Ser	Pro	Val	Thr	Cys	L/I	Asn	Asp	K/Q	Glu	Met	His	Phe	Arg	CmC	Tyr	Trp
Gly	114																		
Ala	128	142																	
Ser	144	158	174																
Pro	154	168	184	194															
Val	156	170	186	196	198														
Thr	158	172	188	198	200	202													
Cys	160	174	190	200	202	204	206												
L/I	170	184	200	210	212	214	216	226											
Asn	171	185	201	211	213	215	217	227	228										
Asp	172	186	202	212	214	216	218	228	229	230									
K/Q	185	199	215	225	227	229	231	241	242	243	256								
Glu	186	200	216	226	228	230	232	242	243	244	257	258							
Met	188	202	218	228	230	232	234	244	245	246	259	260	262						
His	194	208	224	234	236	238	240	250	251	252	265	266	268	274					
Phe	204	218	234	244	246	248	250	260	261	262	275	276	278	284	294				
Arg	213	227	243	253	255	257	259	269	270	271	284	285	287	293	303	312			
CmC	218	232	248	258	260	262	264	274	275	276	289	290	292	298	308	317	322		
Tyr	220	234	250	260	262	264	266	276	277	278	291	292	294	300	310	319	326	326	
Trp	243	257	273	283	285	287	289	299	300	301	314	315	317	323	333	342	349	349	272

corresponds to the loss of I/L (mass: 634 − 531 = 113 amu), which we usually write as Lxx (though we cannot rule out GG) and is a y_4 ion (FVQK).

Next, we start looking for a series of ions. If the 521-amu ion is really a y_4 ion, there may be a weak b_1 ion with a mass of X, where 521 + X = 634 + 1; X is 114 amu, which is clearly present. The next major ion after the 521-amu peak has a mass of 374 amu and seems to lose 17 amu easily; it also becomes monomethylated and is therefore consistent with a y-type ion corresponding to a mass of 521 − 374 = 147 amu, i.e. Phe. The corresponding b-type ion would have a mass of X, where 374 + X = 634 + 1; X is 261 amu, which agrees with sequence LF. Returning to the y-ion series, the next major ion with a mass less than 374 amu is 275 or 261 amu, indicating a loss of 99 amu (V) or 113 amu (Lxx). The ion with a mass of 261 amu exhibits the loss of 28 amu, indicating an a/b pair; it is not methylated, making it even less probable. The corresponding b-type ion (mass = 372 amu) is not present. However, the b_3 ion corresponding to a y_2 of mass 275 amu has a mass of 360 amu, which is clearly present. Since we are assuming that the C-terminal residue is Lys, then a mass of 128 amu (275 − 147) indicates Gln (or AG). This would explain the loss of 17 amu (ammonia) from the other y-type ions and makes the assignment AG/GA unlikely. Thus, the sequence appears to be Lxx–Phe–Val–Gln–Lys. The immonium ions with masses of 86 (Lxx), 120 (Phe) and 72 amu (Val) are consistent with this composition. This is consistent with the acetylation data, which indicate that two acetyl groups are added, one to the free N-terminus of the peptide and the second to the ε Lys amine group. Repeating the steps with the methylated spectrum confirms the initial reading, and we can account for all the ions present in the spectrum; thus, we can be reasonably confident with the interpretation.

8.2.4 The Importance of Charge Location

The MS/MS spectrum shown in Fig. 8.5, in which the spectrum is dominated by one ion series (the y-type ions) helps to explain why virtually all digestions done preparatory to MS/MS analysis are performed using trypsin. The presence of a fixed charge at each terminus gives a dominant y-type series and a weak b-type series (Sect. 8.1.3). The importance of charge location in a digest fragment is illustrated in the MS/MS spectrum of a chymotryptic fragment shown in Fig. 8.6. The masses of the native, methylated and acetylated peptides are given in the figure (629, 643 and 671 amu, respectively). Since these are all singly charged ions, there is no need to deconvolve the spectrum. Comparing the native and methylated spectra, we see that none of the daughter ions increases its mass by 14 amu, which indicates that, because this is a chymotryptic peptide, there are no acidic residues in the sequence, and all the ions form a b-type series. There are two ammonia (17-amu) losses from the 266- and 365-amu ions. At the low-mass end, there is a strong signal at 110 amu,

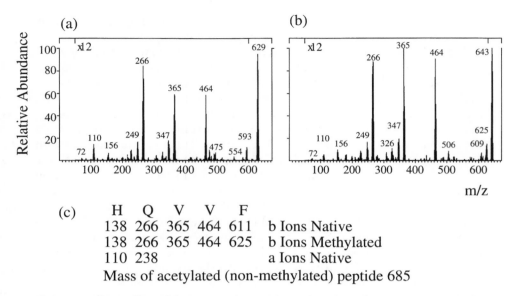

(c)   H   Q   V   V   F
       138 266 365 464 611   b Ions Native
       138 266 365 464 625   b Ions Methylated
       110 238   a Ions Native
    Mass of acetylated (non-methylated) peptide 685

Fig. 8.6a–c. Charge-directed fragmentation. A triple-quadrupole tandem mass spectrum of a chymotryptic peptide with an N-terminal His in the **a** native and **b** methylated states. **c** The sequence, together with the most important sequence ions

indicating the possible presence of His. There are no clear a/b pairs and, again, there appears to be a loss of 17 amu from some ions.

The first major ion at the high-mass end has a mass of 629 − 464 = 165 amu, which does not correspond to any y-ion residue mass; however, (629 − 18) − 464 = 147 amu, which is the mass of a b-type ion (Phe), which is supported by the presence of an immonium ion of mass 120 amu. There appears to be no corresponding y-type ion. The next major ions have masses of 365 and 347 amu, indicating the loss of Val, because 347 amu does not correspond to a residue mass but to the loss of ammonia from a 365-amu ion. The next ion has a mass of 365 − 266 = 99 amu (Val). Again, there is no corresponding y-type ion. The 266-amu ion loses ammonia to give the 249-amu ion. The missing ion is the b-type ion of His, which is not present, but the a-type ion equivalent (mass = 110 amu) is present. The missing residue has a mass of 266 − (110 + 28) = 128 amu (Gln), which explains the extensive losses of ammonia throughout the spectrum. This interpretation is consistent with all the low-mass ions observed, the methylated MS/MS spectrum and the single addition of an acetyl group to the N-terminal (as shown by the acetylated-peptide mass).

The major difference between the MS/MS spectra of the tryptic peptide with the N-terminal Lys (dominated by y-type ions) and that of chymotryptic peptide with the N-terminal His (which has only b-type ions) is the position of the charge-carrying group. The guiding principle is: find the charge carrier; the spectrum is then self-explanatory. This is especially true when the charge carrier is found in the middle of the sequence (a partial tryptic fragment or a standard AspN-digest product). One must be sure that *all* ions are accounted for; otherwise, the results can be deceptive. One problem that may occur is that it may be impossible to decide the order of the first two amino acids. There are two solutions to this: one either performs one cycle of manual Edman degradation (Hunt et al. 1986) or derivatises the N-terminus with a specific reagent and performs MS/MS (for acetylates or nicotinylates; Sect. 8.3.2). A final difficulty is the interpretation of MS/MS spectra of post-translationally modified proteins, which we will not discuss here, since it is discussed in Chapters 10 and 11.

8.3 Aids to Spectral Interpretation and the Avoidance of Pitfalls

8.3.1 Distinguishing Between Residues with Closely Related Masses

Low- and high-energy MS/MS fragmentation spectra do not always give unequivocal solutions to the peptide sequence. Often, an additional step is needed to resolve ambiguities. The two triple-quadrupole examples demonstrated the use of methylation (to help identify the y-type ion series and to avoid mis-assignments of Glu/Gln and Asp/Asn) and acetylation (to resolve Lys/Gln). These problems have been largely overcome with Q-TOF instruments, where mass accuracy and resolution of the daughter ions is sufficient to overcome the problems of quasi-isobaric amino acid combinations. The true isobaric pair Ile/Leu cannot be distinguished, though this may be resolved in the future, using techniques like surface-induced dissociation (Sect. 8.1.2).

One regularly occurring problem is the low cleavage often observed at Gly–Gly and Gly–Ala bonds, which leads to confusion, because the mass of Gly–Gly is equal to that of Asn, and that of Gly–Ala is equal to that of Glu. The latter pair is easily resolved by methylation. One further isobaric pair is Val–Gly and Arg. Methylation does not help here, and there are two possibilities if there are no telltale losses of ammonia from Arg (or if one cannot infer Arg from its C-terminal position after a tryptic digestion). One can modify Arg with a site-specific reagent, such as a dione like diacetylacetone or hydroxyphenylglyoxal (Takahashi 1968; Yankeelov et al. 1968). However, these reagents do not yield the cleanest results and are not efficient when dealing with sub-picomole amounts of material. A general solution to the problem is the use of deuteration (Table 8.1). This has the advantage of occurring very rapidly, with almost perfect yields, though it is technically difficult to analyse the peptides afterwards without a certain degree of back-exchange (Spengler et al. 1993). This method can distinguish between all mono- and di-residue substitution pairs and has the added advantage of allowing a rapid comparison between native and deuterated MS/MS spectra, because there are no major changes in the pattern (as occurs with methylation or acetylation; Sepetov et al. 1993).

8.3.2 Isotopic Labelling of Ion Series

One aid to interpretation, isotopic labelling of the peptide to allow the b- and y-type ion series to be distinguished, has been in use for a long time in MS/MS. A 50:50 mixture of $CD_3COOH:CH_3COOH$ as the acid anhydride was used to acetylate peptides at a pH of 5 (thus, one specifically labels the N-terminus and not the ε amines of Lys; Hunt et al. 1981). This method suffers from two drawbacks: (1) the acetylation causes a drop in the ionisation efficiency, so approximately 20 times more material is needed to obtain an MS/MS spectrum and (2) the labelling specificity is variable and, often, modification of Lys and Tyr residues is observed, negating the usefulness of ion-series labelling. Other non-isotopic labels have been suggested; these involve placing a fixed positive charge on the N-terminal residue (Stults et al. 1993; Cardenas et al. 1997; Spengler et al. 1997; Huang et al. 1999). This increases the ionisation efficiency and, hence, the sensitivity and greatly increases the intensities of all the b-type ions. We have developed a similar approach, which combines isotopic and fixed charge labelling using nicotinic acid (Fig. 8.7).

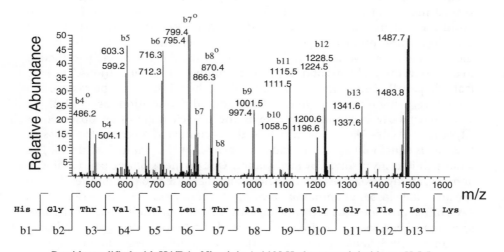

Fig. 8.7. Isotopic labelling of ion series. A tryptic peptide was modified with a 50:50 molar mixture of H4/D4–nicotinic acid *N*-hydroxysuccinimide at pH 5, which specifically labels the N-terminal. *Above* The ion-trap tandem mass spectrum of the isotope clusters 1483 (non-deuterated) and 1487 of the parent ion. All the b-type ions appear as doublets separated by 4 amu, because they contain the modified N-terminus, whilst all y-type ions appear as singlets. *Centre* The fragmentation scheme of the peptide. *Below* The N-terminal modifying group

One of the main problems in obtaining a complete protein digest is the difficulty of denaturing the protein completely without affecting the endopeptidase. We have returned to the old method of succinylating the protein (Klotz 1967) completely at the Lys residues. This is a very clean reaction with few side reactions. The protein is then digested with trypsin and is modified with nicotinoyl-N-hydroxysuccinic acid (50% D4, 50% H4) at a pH of 8.5 to label all the N-termini. Thus, all the peptides appear as a pair of peaks separated by 4 amu. Since the peptide is labelled with a permanent positive charge on the N-terminus, the MS/MS spectrum is dominated by b-type ions, which appear as doublets; the y-type ions appear as single-isotope clusters (Fig. 8.7). Thus, the sequence can be rapidly and easily read and, because all the reagents are water soluble, the reactions can be performed with very small amounts of peptide immobilised on reverse-phase columns and washed prior to analysis. Another advantage of this method is that it allows the order of the first two amino acids to be determined easily, especially if an ion-trap mass spectrometer is being used; the smallest b-type ion can be immediately located and isolated for MS/MS/MS analysis.

8.3.3 A Cautionary Tale

Currently, the trend is the analysis of protein digests by static nanospray using a nanospray tip filled with a microlitre of digest and the subsequent isolation of ions for analysis by the mass spectrometer. However, care should be exercised because, when using an entire unseperated digest, several peptides with the same m/z may be selected and yield uninterpretable, multiple superimposed mass spectra. An example of this is shown in Fig. 8.8, in which a singly charged predominant ion overlaps with a weaker, doubly charged ion. This was immediately obvious, because ions with masses greater than the singly charged parent's mass were observed. In this particular case, which is rather simple, one uses the masses greater than that of the parent ion (which can only result from the one doubly charged peptide) to determine a sequence tag (Sect. 8.4) with which to identify the peptide. The theoretical masses of this peptide are then removed from the spectrum, and a second tag search can be carried out. Therefore, we strongly recommend using high-performance liquid chromatography (HPLC) or capillary zone electrophoresis (CZE) on-line pre-separations (very fast run times can be obtained with both methods, and data accumulation/searching can also be automated).

8.4 Manual Interpretation and Database Searching

Even using the modifications described above to label the ions as b- or y-type series, greatly simplifying interpretation, there will always be peptides that do not give spectra that can give a full sequence. There are often gaps in the

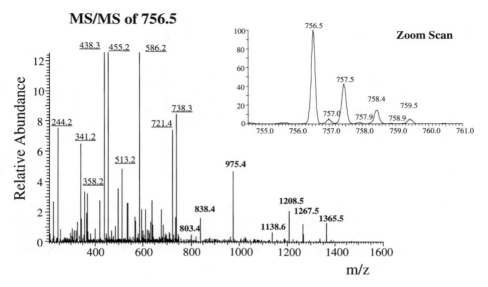

Fig. 8.8. Tandem mass spectrometry (MS/MS) of overlapping peptides in static nanospray. Two peptides can be observed in the zoom scan shown in the *inset*: (1) a singly charged peptide at 756.5 amu, corresponding to the sequence GLMPNPK from the *Escherichia coli* protein 50S subunit L1. The ions from this peptide are *underlined* in the MS/MS spectrum. (2) A doubly charged peptide at 758.0 amu, corresponding to the sequence FRPGTDEGDYQVK from *Escherichia coli* protein-chain initiation factor IF-3. The corresponding daughter ions in the MS/MS spectrum are marked in *bold*

spectra and regions that give equivocal results (GG sequences, very Pro- and His-rich peptides, etc.). However, one of the limiting factors in interpreting MS/MS spectra is the time required to derive the sequence. In a single automated MS/MS HPLC or CZE run lasting 30 min, over 200 MS/MS spectra can be accumulated. One solution to this problem is to use the MS/MS spectrum as a "peptide-fragment fingerprint" and to use it to search a sequence database in a manner analogous to peptide-mass fingerprinting. The group of John Yates developed the SEQUEST program, which can fully automate database searching and is described in detail in Chapter 7.

An alternative approach was proposed by Matthias Mann's group and was developed as the PeptideSearch program (Mann and Wilm 1994). The algorithm developed for database searching, PeptideSearch, requires the intact peptide mass, the tag sequence and the tag sequence's start and end masses (Fig. 8.9, below). In cases of an amino acid difference due to a mutation, a DNA-sequencing mistake or a post-translational modification, the method can tolerate the mass change between the actual peptide and the one predicted from the database. The authors stressed the concept that, for successful protein identification, even a small sequence of two or three amino acids from a noisy spectrum or a partial fragmentation can be used. The method appears to be

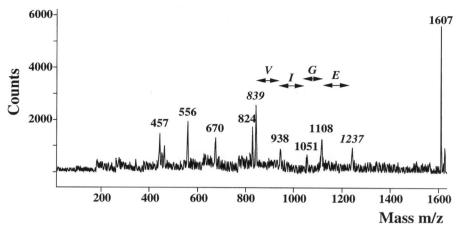

PeptideSearch match using all 3 regions, (839)VIGE(1237) 1607 +/-2

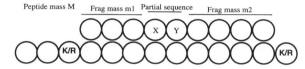

Fig. 8.9. PeptideSearch tag construction. *Above* The post-source decay matrix-assisted laser desorption ionisation time-of-flight spectrum of a tryptic peptide with a mass of 1607 amu. *Below* The format of the tag used for database searching: m1(sequence)m2, parent mass. The PeptideSearch algorithm correctly identified the protein as GroEL3 (*Bradyrhizobium japonicum*) using the tag "(839)VIGE(1237) 1607" with a mass accuracy of ±2

especially useful for the analysis of protein mixtures and the assignment of the proteolytic fragments to the different components.

Lennon and Walsh (1997) showed that, with delayed extraction on a linear TOF mass spectrometer, it was possible to study the in-source decay of metastable, MALDI-generated ions. The method mainly provides a set of c_n ions, and the spectra show fragmentation six or seven residues from either end of the polypeptide chain. In linear mode, the entire mass range of metastable-decay ions can be accumulated in a single spectrum, allowing the direct sequencing of intact proteins without the need for prior digestion. This method has only been reported for known proteins with a mass less than 18 kDa and with amounts greater than 10 pmol.

More recently, with the development of ion-series tagging methods (especially the revival of $^{18}O/^{16}O$ labelling; Takao et al. 1991; Schnolzer et al. 1996), it has been possible to automate the extraction of the tag data for searching (Shevchenko et al. 1997). One of the strengths of the method is that even poor spectra with few daughter ions can yield good tags. Figure 8.9 (above) shows a post-source decay spectrum of a peptide; the spectrum shows a continuous set of ions from which a sequence tag could be extracted and successfully used to identify the peptide in the database. The program allows one to relax the constraints regarding which parameters are fixed. For example, one can require that only M_1 and the sequence tag should fit the data, allowing post-translationally modified peptides to be identified (even when the mass and nature of the modification is undefined). This also allows one to search for homologous peptides to a certain degree. One caveat should be mentioned: multiple MS/MS spectra should be used in order to identify a protein with a high degree of confidence (Chap. 7). An alternative approach that should also profit from ion-series labelling techniques is the automation of spectral interpretation, which is discussed in Chapter 9.

8.5 Conclusion

Despite the fact that MS/MS data accumulation and database searching is fully automatable and is becoming almost routine (Piccinni et al. 1994; Stahl et al. 1996; Figeys et al. 1999; Tong et al. 1999), there will always remain a need for manual confirmation of the database-match assignment. The most obvious case where this will remain necessary is post-translationally modified peptides. There are over 300 naturally occurring modifications, and the identification of their nature and sites of binding require manual interpretation. The development of software tools, such as Sherpa (Taylor et al. 1996) and the SEQUEST helper programs, can greatly ease matters, but a basic knowledge of the process of interpretation is necessary. Certain peptides, especially those with no defined fixed charge (for example, peptides derived from the major histocomaptibility complex), give spectra that do not yield ion series that are easy to predict, and must be confirmed manually (Hunt et al. 1992). In fact, the ultimate proof of a sequence assignment is the synthesis of the peptide and a direct comparison of the spectra from the in vivo and synthetic analogues. Recent advances in computation may allow the development of "intelligent" algorithms that can be trained with the large data sets that are being accumulated. These will be able to extract and apply rules and may eventually be able to match human interpretation in the same way that chess programs have evolved to be competitive with their masters.

Acknowledgements

Both authors attended the course on MS/MS interpretation that has been organised yearly by Prof. Don Hunt in Charlottesville, Virginia, since 1989. That the authors were heavily influenced by this pioneering course is in no small part due to the excellence of the teaching and the thoroughness of the introduction to this art (and, to a large extent, to the many long nights puzzling over the multitude of examples). This course has influenced many of the current protein chemists turned peptide MS/MS practitioners. We extend our thanks to Prof. Don Hunt for the chance to learn from the godfather (Nomen ist Omen) of peptide MS/MS.

References

Barber M, Bordoli RS, Sedgwick RD, Taylor AN (1981) Fast atom bombardment of solids as an ion source in mass spectrometry. Nature 293:270–275

Biemann K (1990) Appendix 5. Nomenclature for peptide fragment ions (positive ions), Methods Enzymol 193:886–887

Cardenas MS, van der Heeft E, de Jong APJM (1997) On-line derivatisation of peptides for improved sequence analysis by micro-column liquid chromatography coupled with electrospray ionisation tandem mass spectrometry. Rapid Commun Mass Spectrom 11:1271–1278

Fenn JB, Mann M, Meng CK, Wong SF, Whitehouse CM (1989) Electrospray ionization for mass spectrometry of large biomolecules. Science 246:64–71

Figeys D, Corthals GL, Gallis B, Goodlett DR, Ducret A, Corson MA, Aebersold R (1999) Data-dependent modulation of solid-phase extraction capillary electrophoresis for the analysis of complex peptide and phosphopeptide mixtures by tandem mass spectrometry: application to endothelial nitric oxide synthase. Anal Chem 71:2279–2287

Gross ML (1990) Tandem mass spectrometry: multisector magnetic instruments. Methods Enzymol 193:131–153

Huang ZH, Shen T, Wu J, Gage DA, Watson JT (1999) Protein sequencing by matrix-assisted laser desorption ionization – postsource decay-mass spectrometry analysis of the N-Tris(2,4,6-trimethoxyphenyl)phosphine-acetylated tryptic digests. Anal Biochem 268:305–317

Hunt DF, Buko AM, Ballard JM, Shabanowitz J, Giordani AB (1981) Sequence analysis of polypeptides by collision activated dissociation on a triple-quadrupole mass spectrometer. Biomed Mass Spectrom 8:397–408

Hunt DF, Yates JR III, Shabanowitz J, Winston S, Hauer CR (1986) Protein sequencing by tandem mass spectrometry. Proc Natl Acad Sci USA 83:6233–6237

Hunt DF, Henderson RA, Shabanowitz J, Sakaguchi K, Michel H, Sevilir N, Cox AL, Appella E, Engelhard VH (1992) Characterization of peptides bound to the class I MHC molecule HLA-A2.1 by mass spectrometry. Science 255:1261–1263

Johnson RS, Martin SA, Biemann K, Stults JT, Watson JT (1987) Novel fragmentation process of peptides by collision-induced decomposition in a tandem mass spectrometer: differentiation of leucine and isoleucine. Anal Chem 59:2621–2625

Johnson RS, Martin SA, Biemann K (1988) Collision induced fragmentation of MH+ ions of peptides. Side chain specific fragmentation ions. Int J Mass Spectrom Ion Proc 86:137–154

Karas M, Hillenkamp F (1988) Laser desorption ionization of proteins with molecular masses exceeding 10,000 daltons. Anal Chem 60:2299–2301

Klotz IM (1967) Succinylation. Methods Enzymol 11:576–591

Lennon JJ, Walsh KA (1997) Direct sequence analysis of proteins by in-source fragmentation during delayed ion extraction. Protein Sci 6:2446–2453

Mabud MA, DeKrey MJ, Cooks RG (1985) Surface induced dissociation of molecular ions. Int J Mass Spectrom Ion Proc 67:285–291

Mann M, Wilm M (1994) Error-tolerant identification of peptides in sequence databases by peptide sequence tags. Anal Chem 66:4390–4399

McLuckey SA, Van Berkel GJ, Goeringer DE, Glish GL (1994) Ion trap mass spectrometry. Using high-pressure ionization. Anal Chem 66:737A–743A

Morris HR, Paxton T, Dell A, Langhorne J, Bordoli RS, Hoyes J, Bateman RH (1996) High sensitivity collisionally-activated decomposition tandem mass spectrometry on a a novel quadrupole/orthogonal-acceleration time-of-flight mass spectrometer. Rapid Commun Mass Spectrom 10:889–896

Papayannopoulos IA (1995) The interpretation of collision induced dissociation tandem mass spectra of peptides. Mass Spectrom Rev 14:49–73

Piccinni E, Staudenmann W, Albergoni V, De Gabrieli R, James P (1994) Purification and primary structure of metallothioneins induced by cadmium in the protists *Tetrahymena pigmentosa* and *Tetrahymena pyriformis*. Eur J Biochem 226:853–859

Roepstorff P, Fohlman J (1984) Proposal for a common nomenclature for sequence ions in mass spectra of peptides. Biomed Mass Spectrom 11:601

Schnolzer M, Jedrzejewski P, Lehmann WD (1996) Protease-catalyzed incorporation of ^{18}O into peptide fragments and its application for protein sequencing by electrospray and matrix-assisted laser desorption/ionization mass spectrometry. Electrophoresis 17:945–953

Sepetov N, Issakova OL, Lebl M, Swiderek K, Stahl DC, Lee, TD (1993) The use of hydrogen-deuterium exchange to facilitate peptide sequencing by electrospray tandem mass spectrometry. Rapid Commun Mass Spectrom 7:58–62

Shevchenko A, Chernushevich I, Ens W, Standing KG, Thomson B, Wilm M, Mann M (1997) Rapid 'de novo' peptide sequencing by a combination of nanoelectrospray, isotopic labeling and a quadrupole/time-of-flight mass spectrometer. Rapid Commun Mass Spectrom 11:1015–1024

Spengler B, Lutzenkirchen F, Kaufmann R (1993) On-target deuteration for peptide sequencing by laser mass spectrometry. Org Mass Spectrom 28:1482–1490

Spengler B, Luetzenkirchen F, Metzer S, Chaurand P, Kaufmann R, Jeffrey W, Bartlet-Jones M, Pappin DJC (1997) Peptide sequencing of charged derivatives by post-source decay MALDI mass spectrometry. Int J Mass Spectrom 169/170:127–140

Stahl DC, Swiderek KM, Davis MT, Lee TD (1996) Data controlled automation of liquid chromatography tandem mass spectrometry analysis of peptide mixtures. J Am Soc Mass Spectrom 7:532–540

Stults JT, Lai J, McCune S, Wetzel R (1993) Simplification of high-energy collision spectra of peptides by amino-terminal derivatization. Anal Chem 65:1703–1708

Takao T, Hori H, Okamoto K, Harada A, Kamachi M, Shimonishi Y (1991) Facile assignment of sequence ions of a peptide labelled with 18O at the carboxyl terminus. Rapid Commun Mass Spectrom 5:312–315

Takahashi K (1968) The reaction of phenylglyoxal with arginine residues in proteins. J Biol Chem 243:6171–6179

Taylor JA, Walsh KA, Johnson RS (1996) Sherpa: a Macintosh-based expert system for the interpretation of electrospray ionization LC/MS and MS/MS data from protein digests. Rapid Commun Mass Spectrom 10:679–687

Tong W, Link A, Eng JK, Yates JR III (1999) Identification of proteins in complexes by solid-phase microextraction/multistep elution/capillary electrophoresis/tandem mass spectrometry. Anal Chem 71:2270–2278

Vachet RW, Glish GL (1998) Boundary-activated dissociation of peptide ions in a quadrupole ion trap. Anal Chem 70:340–346

Yankeelov JA, Mitchell CD, Crawford TH (1968) A simple trimerisation of 2,3-butadione yielding a selective reagent for the modification of arginine in proteins. J Am Chem Soc 90:1664–1668

Yost RA, Boyd RK (1990) Tandem mass spectrometry: quadrupole and hybrid instruments. Methods Enzymol 193:154–200

9 Automated Interpretation of Peptide Tandem Mass Spectra and Homology Searching

RICHARD S. JOHNSON

9.1 Database-Searching Programs for Inexact Peptide Matches

Computer programs that match tandem mass spectrometry (MS/MS) data to database sequences typically search first for exact matches between the observed peptide mass and the mass calculated from the database (Eng et al. 1994). If exact matches cannot be found, some of these programs have the ability to find database sequences that do not match the observed peptide mass. The program PeptideSearch (Mann and Wilm 1994) uses the concept of a "peptide-sequence tag", which is a partial sequence containing an unsequenced mass on the C- and N-terminal ends of a short stretch of sequence. For triple-quadrupole and quadrupole/time-of-flight (QTOF) data of tryptic peptides, the short stretch of sequence is often easily determined from the prominent series of y-type ions that usually is present at masses greater than those of multiply charged precursor ions. Sequence tags are more difficult to obtain from ion-trap data, where high-mass b- and y-type ions are both typically present. Since peptide-sequence tags contain three parts – the N-terminal unsequenced mass, the C-terminal unsequenced mass and the intervening short sequence – it is possible to search a sequence database, requiring an exact match for only two of the three regions. Of course, by reducing the search constraints in this manner, the program is much more likely to find false positives, and it has been the author's experience that several tens of possibilities are obtained. It is difficult but necessary to manually sift through the numerous sequence candidates to decide which (if any) are likely to be correct matches. Also, homologous substitutions are not given any consideration when deriving the sequence candidates.

MS-Tag (Clauser et al. 1996) is another database-search program that is commonly used in "identity mode" for finding exact database matches using MS/MS data, but it can also be used in "homology mode" to find database sequences that have calculated peptide masses that differ from the observed mass. In identity mode, this program finds database sequences that match the observed peptide mass, then determines whether any of these candidate sequences can account for the input fragment ions. From the peptide mass, it is possible to calculate the neutral segment of the peptide that was lost during fragmentation to produce the observed fragment ion. Both pieces of the peptide represented by the fragment ion and the resulting neutral fragment must agree with the sequence obtained from the database.

In homology mode, the database is searched for peptide sequences that are close to the observed mass (±45 u), and the program only requires that either the fragment ion or the neutral fragment (not both) have a mass that agrees with the database sequence. By assuming that the difference in mass compared with the database sequence is due to a single amino acid change, the program can utilise a "mutation matrix" that ranks the sequences according to the most homologous amino acid change that can account for the mass shift. If more than one amino acid is different, or if a single mutation alters the proteolytic cleavage site that is not present in the peptide (the lysine or arginine that releases the N-terminus of a tryptic peptide), then this approach will fail. An alternative means of making inexact matches to database sequences is to first deduce an amino acid sequence from the MS/MS spectrum without the aid of a sequence database, then use the deduced sequence to find homologous matches using a sequence-database search program [FASTA (Pearson and Lipman 1988) or Blast (Altschul et al. 1990)].

9.2 Algorithms for Automated Interpretation of Peptide Tandem Mass Spectra

Automated peptide-sequence determinations from MS/MS spectra have had a chequered past in which a variety of algorithms have been proposed; none have enjoyed widespread use or success. From a single MS/MS spectrum, it is impossible to deduce a sequence that one is completely certain is correct. However, imperfect and partially correct sequences can still be extremely useful; furthermore, by using a variety of techniques, it is possible to increase the reliability of a sequence.

9.2.1 Brute-Force Method

Conceptually, the simplest method for interpreting peptide MS/MS spectra involves a brute-force approach in which all possible sequences that match a measured peptide mass are scored and ranked according to how well the candidate sequences account for the fragment ions in the MS/MS spectrum (Sakurai et al. 1984; Hamm et al. 1986). Since the number of possible sequences grows exponentially with sequence length, excessive processing limits the utility of this approach for larger peptides.

9.2.2 Sub-Sequencing Approach

A second algorithm for sequencing peptides from mass spectra, referred to as "sub-sequencing", was initially used to interpret electron-impact ionisation mass spectra of peptide derivatives (Biemann et al. 1966). This approach has also been applied to sequence peptides using fast-atom-bombardment ionisation (Ishikawa and Niwa 1986; Siegel and Bauman 1988), high-energy collision-induced dissociation (CID; Johnson and Biemann 1989) and low-energy CID mass spectra (Johnson et al. 1991; Yates et al. 1991). In this method, all possible amino acids are considered for placement in one of the terminal positions (either the C- or N-terminus). Amino acid residues that can account for some of the observed fragment ions are retained as candidate subsequences that are sequentially extended by one more amino acid. At each stage of subsequence extension, the sequences are tested against the MS/MS spectrum, and those that account for the most fragment ions are lengthened until they match the observed peptide mass. One of the problems with this type of algorithm is that some peptides have sections that are poorly represented by any fragment ions. During the sequential addition of amino acids, the correct partial sequence may be eliminated before a segment of the MS/MS spectrum that has a greater abundance of sequence-specific fragment ions is reached.

9.2.3 Graphical Display to Assist Manual Interpretation

A third type of computer program makes use of an interpreter's visual skills at detecting patterns by providing a graphical display that highlights fragment ions that differ in mass by an amino acid residue in an MS/MS spectrum (Scoble et al. 1987). In such a display, it is easier to spot a series of ions of the same type and to quickly postulate a sequence that accounts for the observed fragment ions. Of course, this is really a computer-assisted manual interpretation and is not suitable for high-throughput data interpretation. Nevertheless, such programs are useful for cases that require careful manual interpretation.

9.2.4 Graph-Theory Approach

The final category of algorithms for interpreting MS/MS spectra of peptides employs ideas obtained from the mathematical field of graph theory. It was first proposed by Bartels (1990) and later by Hines et al. (1992) and Fernandez-de-Cossio et al. (1995) for high energy CID spectra. The program Lutefisk (Taylor and Johnson 1997) was originally designed to work with low-energy CID spectra obtained from triple-quadrupole mass spectrometers, but it has since been modified to utilise data obtained from ion traps and QTOF instruments.

In the graph-theory approach, each ion is assumed to be one of the allowed fragment ion types, e.g. b- and y-type ions are the main sequence-specific ions in low-energy MS/MS spectra (Papayannopoulos 1995). Each assumed ion type is mapped to a graph of a single ion type. To illustrate the point, Fig. 9.1a shows a hypothetical MS/MS spectrum of the peptide Ala–Ala–Lys, containing ions

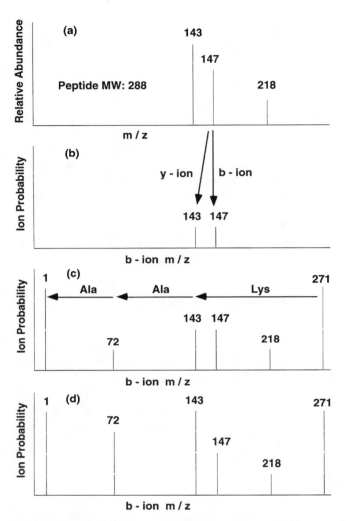

Fig. 9.1a–d. Example illustrating the use of a b-type ion graph for peptide sequencing. **a** A hypothetical mass spectrum of the peptide Ala–Ala–Lys, with a molecular weight of 288 u. The ions at m/z = 143, 147 and 218 u are b_2-, y_1- and y_2-type ions, respectively. **b** The b-type ion graph after assuming the ion with an m/z of 147 is either a b- or y-type ion. **c** The b-type ion graph after assuming that all of the ions (m/z = 143, 147 and 218) are b- or y-type ions. The node positions at 1 and 271 correspond to the N- and C-termini of the peptide, respectively. **d** The node positions 72 and 143 can be connected to the C-terminal node at 271 via single residue masses and are assigned a higher ion probability value

at m/z = 143, 147 and 218. If the ion with an m/z of 147 is assumed to be a b-type ion, it maps to the b-type ion graph in Fig. 9.1b as a b-type ion at position (node) 147. The node 147 in Fig. 9.1b is assigned an "ion probability" value for b-type ions, which is a value that estimates the importance of b-type ions in determining a peptide sequence relative to other ion types. In addition to possibly being a b-type ion, the ion with an m/z of 147 could also be a y-type ion, which can be converted to its corresponding b-type m/z if the mass of the intact peptide is known. Using the formula b-type ion = (peptide mass) − (y-type ion mass) + 2 × (mass of hydrogen), the ion with an m/z of 147 can be converted to its corresponding b-type ion with an m/z of 143, and the y-type ion "ion probability" is placed at node 143 (Fig. 9.1b). Likewise, the ion with an m/z of 218 (Fig. 9.1a) is assumed to be both a b-type ion and a y-type ion, which map to nodes 218 and 72 on the b-type ion graph, respectively (Fig. 9.1c). The ion with an m/z of 143 (Fig. 9.1a) also maps to nodes 143 and 147 in the b-type ion graph, and "ion probabilities" from the ion with an m/z of 143 are added to the probabilities derived from the ion with an m/z of 147 (Fig. 9.1c). Of course, the b-type ion graph has to have end points corresponding to the N- and C-termini of the peptide. For unmodified peptides, the N-terminal group is a proton; hence, node 1 is assigned an arbitrary, non-zero positive value (Fig. 9.1c). The C-terminal node in the b-type ion graph can be calculated from the peptide mass minus the mass of the C-terminal group (a hydroxyl, in the case of an unmodified C-terminus). Hence, 271 (288–17) represents the C-terminus of the peptide in the b-type ion graph (Fig. 9.1c). As an alternative to assuming that each ion is one of the allowed ion types, Scarberry et al. (1995) proposed using a neural network to first classify fragment ions from high-energy CID spectra into specific ion types prior to forming the graph.

One of the advantages of creating the b-type ion graph is that it is possible to first identify the nodes that can be connected to the C-terminal node, e.g. 271 in Fig. 9.1c, via connections (edges) corresponding to single amino-acid-residue masses. Such nodes are then given a higher probability, because they represent pathways through the graph that could lead to a complete, uninterrupted sequence. As shown in Fig. 9.1c, only nodes 143 and 72 can be connected to the C-terminal node 271 by their edges, and these nodes are assigned higher probability scores (Fig. 9.1d). Thus, the magnitude of the probability (the y-axis in Fig. 9.1b–d) for each node contains information regarding the presence of N- and C-terminal fragmentation, plus the likelihood of reaching the C-terminus. From the graph shown in Fig. 9.1d, sequences are determined by starting at the N-terminal node (1) and finding all of the edges that can lead to the C-terminal node (271). Although only one sequence is possible for this simple example, data from larger peptides can result in graphs that produce many thousands of sequences. Therefore, in order to complete the sequencing in a reasonable time, a limit is placed on the number of N-terminal sub-sequences that are tracked. This is accomplished by using only those subsequences that utilise nodes with the highest probability. Sequences encompassing low-probability nodes are discarded.

Several thousand completed sequences can be generated by this approach. Lutefisk (Taylor and Johnson 1997) sorts through this list by first eliminating sequences that lack continuous series of b- and y-type ions. Tryptic peptide sequences that were derived from alternating b- and y-type ions are not likely to be correct. In addition, the region of the spectrum above the precursor ion (for multiply charged ions) generally contains a few intense ions, which the candidate sequences must be able to explain; those that do not are discarded. These tests usually eliminate more than 90% of the sequence candidates, and the remaining ones are scored and ranked. From a single MS/MS spectrum, the output for Lutefisk is a list of as many as five candidate sequences. Lutefisk has been compiled on UNIX, Macintosh and Windows platforms; the source code and executable files for Macintosh and Windows are available via the internet (http://www.hairyfatguy.com).

9.3 Ambiguities Associated with Sequence Determinations from MS/MS Spectra

The use of MS/MS for peptide-sequence determination was first demonstrated using a triple-quadrupole to analyse peptides of known sequence derived from apolipoprotein B (Hunt et al. 1986) and, shortly thereafter, several small proteins of unknown structure were sequenced using high-energy CID spectra obtained from a four-sector tandem mass spectrometer (Johnson and Biemann 1987; Johnson et al. 1988a,b; Hopper et al. 1989). However, in these and later reports, some of the difficulties in interpreting and extracting an unambiguous sequence from MS/MS spectra were not always mentioned.

9.3.1 Isomeric and Isobaric Amino Acid Residues

Leucine and isoleucine are isomeric amino acids differing only in the position of a side chain methyl group. The differentiation of these amino acids was shown to be possible for high-energy CID using the so-called w-type ions and d-type ions (Johnson et al. 1987); however, the more commonly used low-energy CID spectra do not exhibit these ions. Hence, spectra obtained from triple-quadrupole, ion-trap and QTOF instruments are not able to make this differentiation. Differences in the fragmentation of leucine and isoleucine immonium ions with an m/z of 86 have been described (Hulst and Kientz 1996). In some cases, it may be possible to perform in-source CID to produce the Leu/Ile immonium ion, which can then be subjected to MS/MS. Of course, this requires that the in-source CID be performed on a single peptide and that the peptide does not contain both leucine and isoleucine.

Glutamine and lysine differ in mass by only 0.036 u. QTOF mass spectrometers have sufficient mass accuracy to make this distinction (Morris et al. 1996); however, triple quadrupoles and quadrupole ion traps do not. Since lysine contains a reactive amino group, it is possible to introduce a mass difference between these amino acids by chemical modification (acetylation). By performing two stages of MS/MS on a quadrupole ion trap (MS³), it is possible to distinguish between glutamine and lysine (Bahr et al. 1998). MS³ spectra show that b-type ions containing C-terminal glutamine lose 45 u, whereas b-type ions containing a C-terminal lysine do not. In practice, however, this approach requires that the user decide which of the MS/MS fragment ions are b-type ions that contain a C-terminal Lys/Gln. This can only be done by first making a sequence determination from the MS/MS spectrum.

9.3.2 Amino Acid Residues That Are Isobaric with Dipeptide Residues

Another source of confusion when sequencing peptides by MS/MS is that combinations of some of the lower mass amino acid residues have identical or nearly identical masses of higher-molecular-weight residues. When interpreting a spectrum, it is possible to overlook a cleavage ion between such amino acid pairs (Ala–Gly, Ala–Asp, Gly–Gly, Ser–Val or Gly–Val) and to incorrectly infer the presence of the corresponding single amino acid. Conversely, a spectrum might have an ion series of the same mass as an unrelated ion, so one could incorrectly conclude that a dipeptide is present when the peptide actually contains the single, higher mass amino acid residue.

9.3.3 Missing Fragmentation

A common feature of low-energy CID MS/MS spectra is a lack of fragmentation between the first and second amino acids. Hence, one can only determine the mass of the two N-terminal amino acids and not their sequence. Likewise, fragmentation on the C-terminal side of proline is often absent or is of reduced intensity. Furthermore, spectra of low intensity or quality might be missing some sequence-specific fragmentation, thus resulting in an incomplete sequence when the mass and sequence position of a dipeptide can be postulated but the exact sequence cannot.

9.3.4 Unknown and Unusual Fragmentation Reactions and Artefacts

Although the major types of fragmentation are well established (Papayannopoulos 1995), one can rarely account for every fragment ion in an MS/MS spectrum. One example was the rearrangement observed when C-terminal amino acids of a b-type ion were transferred to the amino-group side chain of lysine (Tang et al. 1993). Subsequent fragmentation of the rearranged structure produced an ion series that could cause confusion during a sequence determination. Other aberrant ions could result from the presence of a contaminant at the same mass as the selected precursor ion (Davis and Lee 1997). This can be a problem, for example, when selecting a precursor ion of very low signal-to-noise ratio, where a large component of the precursor ion includes the chemical background noise. Artefacts can also arise when using an improperly tuned instrument. Such unusual, unknown or artefactual ions can complicate sequence determinations.

9.3.5 Inherent Ambiguity Due to Spectrum Complexity

If b- and y-type ions were the only fragmentation products observed in low-energy CID, it would be fairly simple to correctly deduce peptide sequences from MS/MS spectra. However, in addition to b- and y-type ions, neutral losses of water and ammonia from b- and y-type ions can occur. In addition to these sequence-specific ions that contain either the N- or C-termini, there are internal fragment ions that lack both termini. Such known and expected fragmentations, combined with the complicating factors described above, conspire to make it very difficult to confidently derive a sequence from a single MS/MS spectrum. In general, automated or manual interpretations of peptide MS/MS spectra can only provide multiple, tentative, partially correct sequences.

9.4 Using Ambiguous Peptide Sequences Derived from Tandem Mass Spectra

9.4.1 Automated Validation of Database Matches

Given that a single MS/MS spectrum of a particular peptide is all that one usually has, what can be done with the acquired data? The main advance has been in the use of MS/MS spectra for making sequence-database matches, as described in other chapters of this book. In a test set of 60 MS/MS spectra of known and unknown peptides, three programs – MS-Tag (Clauser et al. 1996), PepFrag (Fenyo et al. 1998) and SEQUEST (Eng et al. 1994) – were compared to see how well they fared with respect to false positives (incorrect matches) and false negatives (missing a protein that is present in the database). In general, the results were similar and, for various reasons, the programs each yielded approximately six false negatives and four false positives. It was clear from this test set that it can be dangerous to take the results of database searches at face value without further validation.

Experiments combining liquid chromatography with MS/MS (LC/MS/MS) can generate hundreds of MS/MS spectra, and it would be quite a task to man-

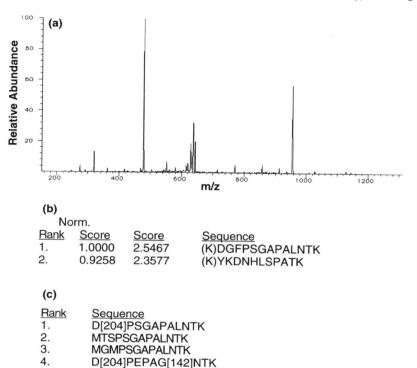

(b)

Rank	Norm. Score	Score	Sequence
1.	1.0000	2.5467	(K)DGFPSGAPALNTK
2.	0.9258	2.3577	(K)YKDNHLSPATK

(c)

Rank	Sequence
1.	D[204]PSGAPALNTK
2.	MTSPSGAPALNTK
3.	MGMPSGAPALNTK
4.	D[204]PEPAG[142]NTK

Fig. 9.2a–c. Use of de novo sequencing to validate a database match. **a** The tandem mass spectrum of the peptide DGFPSGAPALNTK. **b** Results obtained from the database-search program SEQUEST (Eng et al. 1994). **c** Results obtained from the automated de novo sequencing program Lutefisk. (Taylor and Johnson 1997)

ually check each database match. Since Lutefisk (Taylor and Johnson 1997) can perform a de novo interpretation in less time than is required for a database match, it can be used for checking database-search results. An example is shown in Fig. 9.2, where the output from SEQUEST (Eng et al. 1994) is compared with the list of sequences generated by Lutefisk. The criteria for making a database match with SEQUEST are that the peptide score exceed a value of 2.0 and that the second-ranked sequence have a normalised score that is less than 0.9. Using these criteria, no match was found; however, the top-ranked Lutefisk sequence matched the top-ranked SEQUEST sequence. Since the algorithm that Lutefisk used to derive candidate sequences is very different from a database search algorithm, an agreement between the two programs provides automated validation of the database match.

9.4.2 Use of a Homology-Based Search Program That Has Been Modified to Handle Sequence Ambiguities

As described in Section 9.3, the difficulties involved in sequencing peptides using a single MS/MS spectrum result in the generation of multiple candidate sequences, none of which are known to be completely correct. Often, the sequence candidates contain ambiguities, such as unsequenced dipeptide masses (Fig. 9.2b), or replace dipeptides with other single amino acids or dipeptides of the same mass (Sect. 9.3.2). Many of the sequence candidates confuse Ile and Leu and confuse Gln and Lys (Sect. 9.3.1). The question is whether these sequence candidates are correct enough to use in performing a homology-based sequence database search using Blast (Altschul et al. 1990) or FASTA (Pearson and Lipman 1988).

A version of the FASTA program (CIDentify; Taylor and Johnson 1997) that has been modified to more effectively deal with the peculiarities and ambiguities of sequences derived from MS/MS data is available. In addition to using altered scoring tables to account for uncertainties regarding Leu/Ile and Gln/Lys, CIDentify compares multiple query sequences to each database sequence and adds the individual scores to give a score for each database sequence. Other changes to the program relate to mass equivalencies and are summarised in Fig. 9.3.

9.4.2.1 Identifying Peptides Derived from Non-Consensus Cleavages or from Chemical Artefacts

Generally, when performing a sequence-database match using MS/MS spectra, database sequences that are derived from a specific enzymatic cleavage (trypsin cleaves at Lys and Arg) and match the observed peptide mass are found. This is very effective at making identical matches to database sequences; however, if a match cannot be made, the question is whether the peptide came from a novel protein or whether it is a result of something more mundane. Due to long reaction times, contaminating proteases or autolytic degradation, trypsin and other proteases that cleave at specific sites will occasionally cut at non-consensus residues. If a database search using MS/MS spectra is made

Fig. 9.3. Mass equivalencies in CIDentify. In this example, the mass of the dipeptide Gly–Ala in the query sequence is the same as that of Gln in the database sequence. The unsequenced dipeptide mass of 200 u in the query sequence corresponds to the database sequence Val–Thr; the query dipeptide Asn–Thr has the same mass as the database sequence Ser–Lys. Trp in the query sequence has the same mass as Glu–Gly in the database sequence

assuming consensus cleavage, then the match cannot be made. Similarly, an unexpected chemical modification will alter the peptide mass so that it no longer agrees with the mass calculated from the database.

An example is shown in Fig. 9.4, where the peptide had an N-terminal carbamidomethylated cysteine that had cyclised and lost ammonia. The two query sequences obtained from Lutefisk (inset, Fig. 9.4a) were used to obtain a match to fetuin (Fig. 9.4b). In this case, Lutefisk was unaware of the possibility of a cyclised N-terminal carbamidomethylated cysteine; however, it assumed that there was an unsequenced dipeptide of mass 258 u at the N-terminus and correctly determined the rest of the sequence. CIDentify made an exact match to fetuin using the sequence SSPDSAEDVRK, and it was subsequently realised that the dipeptide mass of 258 u corresponded to the mass of a cyclised N-terminal carbamidomethylated cysteine plus aspartic acid. Similar matches have been made to peptides with N-terminal pyroglutamyl residues, deamidation, oxidised methionine and peptides derived from non-consensus proteolytic cleavages.

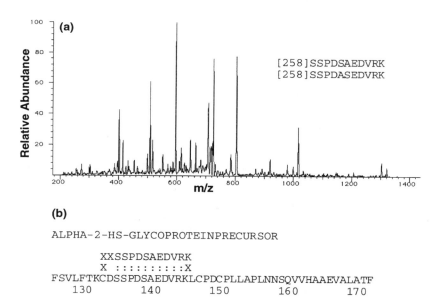

Fig. 9.4a,b. Using Lutefisk and CIDentify to identify a peptide with an unusual chemical modification. **a** The tandem mass spectrum of a tryptic peptide from fetuin. The *inset* shows the two sequences obtained using the automated de novo sequencing program Lutefisk (Taylor and Johnson 1997), where *[258]* at the N-termini denotes the presence of an N-terminal dipeptide of mass 258 u. **b** Using the two sequences from **a**, the homology-based database-search program CIDentify found a close match to a tryptic peptide from fetuin. The mass of the two N-terminal amino acids (carbamidomethylated Cys–Asp) is 17 u higher than the N-terminal dipeptide with mass 258 u found by Lutefisk and is due to the cyclisation of the N-terminal carbamidomethylated Cys and the loss of ammonia

9.4.2.2 Database Sequence Errors and Homologous Proteins

As pointed out in the previous section, an exact match between mass-spectral data and a database sequence may be due to mundane occurrences, such as non-consensus cleavages or chemical artefacts. Another reason for the inability to make an exact match could be inter-species sequence differences, allelic differences or database errors. An example of a homologous protein match is shown in Fig. 9.5, which shows the MS/MS spectrum of a peptide obtained from a mouse protein. No database match was obtained; however, Lutefisk derived a single sequence [(Leu/Ile)–(Lys/Gln–Asn–Val–Asp–Cys–Val–(Leu/Ile)–(Leu/Ile)–Ala–Arg], and CIDentify was able to match it with a human protein that differed at position 7.

9.5 Resolving Sequence Ambiguities

As discussed in Section 9.3, it is nearly impossible to confidently deduce a completely correct sequence from a single MS/MS spectrum; however, by

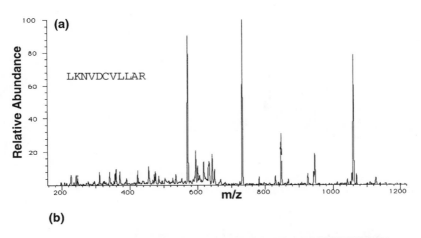

Fig. 9.5a,b. Using Lutefisk and CIDentify to identify a homologous-database sequence. **a** The tandem mass spectrum of a tryptic peptide from mouse methyladenosine phosphorylase. The *inset* shows the single sequence obtained using the automated de novo sequencing program Lutefisk (Taylor and Johnson 1997). **b** Using the sequence from **a**, the homology-based database-search program CIDentify found a close match to a tryptic peptide from human methyladenosine phosphorylase that differs from the mouse protein at position 7

performing a few additional experiments, it is possible to reduce the ambiguity. The general idea is to deduce a sequence from an MS/MS spectrum, then perform additional experiments to try to prove the postulated sequence wrong.

9.5.1 Higher Mass-Accuracy Measurements of Fragment Ions and Peptide Molecular Weights

Just as high mass-accuracy measurements of tryptic peptide masses can improve confidence in protein identification using mass mapping (Henzel et al. 1993), high mass-accuracy measurements of fragment ions in an MS/MS spectrum increase confidence in a sequence assignment. Whereas triple-quadrupole and quadrupole ion-trap instruments will typically produce fragment-ion mass accuracies of $\pm 0.4\,u$, QTOF gives mass accuracies of a few parts per million (Morris et al. 1996). The utility of such high mass accuracy for database searches has been demonstrated (Shevchenko et al. 1997), and is also helpful for de novo sequence determinations using ^{18}O labelling (Sect. 9.5.4). Even without ^{18}O labelling, accurate mass measurements of fragment ions can reduce ambiguities of sequence determinations.

9.5.2 Deuterium Incorporation

By dissolving a peptide in deuterated solvents (D_2O), exchangeable hydrogens in peptides (those attached to oxygen, nitrogen or sulphur atoms) become deuterated. Often, by determining the numbers of exchangeable hydrogens within a peptide, some of the sequence candidates can be eliminated (Sepetov et al. 1993). For example, by measuring the mass of the deuterated peptide, it would be possible to determine that the peptide of Fig. 9.2 has 20 exchangeable hydrogens. Knowing the number of exchangeable protons in this example eliminates two of the sequences in Fig. 9.2c. Further information can be obtained by acquiring the MS/MS spectrum of the deuterated peptide. As described in Section 9.3.2, there are a few amino acid pairs that have masses identical or nearly identical to single amino acids. Deuterium labelling can usually differentiate between these possibilities. For example, Ala–Gly has two exchangeable protons (the two amide hydrogens), whereas Gln/Lys has three (the amide hydrogen plus two side-chain protons). One of the potential difficulties with this approach is the possibility that the peptides will undergo a back-exchange with atmospheric water vapour during the ionisation process. However, this can be controlled by performing the ionisation in a contained ionisation source filled with D_2O vapour or dry nitrogen.

9.5.3 Chemical Modifications

Any covalent modification of a peptide that predictably alters the mass of specific residues can also be helpful for reducing the ambiguities of sequence determinations by MS/MS. The most common procedure in use is to methyl esterify the carboxylates and compare the MS/MS spectra of derivatised and underivatised peptides (Hunt et al. 1986). One of the difficulties with chemical modifications when analysing small quantities of peptide mixtures is that a peptide in its underivatised form may be suppressed or absent when derivatised. Conversely, derivatised peptides that were not seen in the underivatised form may appear; thus, it can sometimes be difficult to obtain derivatised and underivatised pairs of spectra for the same peptide.

9.5.4 C-Terminal ^{18}O Labelling

When a protease cleaves a peptide bond, oxygen from water is incorporated into the newly formed carboxyl ends of the proteolytic peptides. This reaction has been used to isotopically label the C-termini of tryptic peptides (excluding the C-terminus of the protein, which remains unlabeled; Desiderio and Kai 1983; Rose et al. 1983; Gaskell et al. 1988; Takao et al. 1991; Whaley and Caprioli 1991; Schnolzer et al. 1996). By performing the digestion in a mixture of ^{16}O and ^{18}O water, peptides that have two mass-spectral signals differing by 2 u are produced. C-terminal fragment ions in MS/MS spectra of these labelled peptides exhibit a doublet due to the isotopic heterogeneity at the C-terminal oxygen. This information can be used to determine which ions contain the C-terminus and which do not, which can greatly reduce the difficulties and ambiguities of sequencing peptides by MS/MS. C-terminal ^{18}O labelling can be used with any instrument capable of resolving ions that differ by 2 u. Labelling occurs during proteolysis, which is a step that is nearly always performed for protein identification and does not require additional manipulations or the loss of sample. Since the labelling is isotopic and not chemical (Sect. 9.5.3), there is no danger of differentially ionising one form more than the other. Of course, such labelling has the effect of reducing the sensitivity by one half, because the peptide signal is split between the two isotopic forms.

9.5.5 Multiple Stages of MS/MS

A quadrupole ion trap is capable of performing multiple stages of MS/MS. If sufficient time is available (during a nanospray ionisation experiment), a fragment ion from an MS/MS spectrum can be selected for further fragmentation, which makes it possible to obtain additional data to limit the possible sequence candidates (Arnott et al. 1998). Since y-type ions have the same structure as a protonated peptide, selecting a y-type ion for a second stage of MS/MS is conceptually similar to ^{18}O labelling except that the labelling has occurred at the N-terminus. In this case, the mass difference arises due to the removal of N-terminal amino acids, and a comparison of the MS/MS and MS^3 spectra shows y-type ions of the same mass.

Ions that have shifted in the MS^3 spectra compared with the MS/MS spectrum are not C-terminal ions. An example of this is shown in Fig. 9.6, where

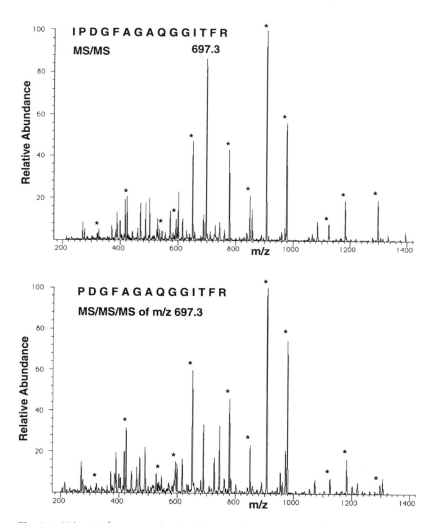

Fig. 9.6. Using MS^3 spectra obtained from an ion trap to verify a sequence deduced from a tandem mass spectrum. *Above* The tandem mass spectrum of the peptide IPDGFAGAQGGITFR. *Below* The MS^3 spectrum of the doubly charged y-type ion with an m/z of 697.3 (A). The *asterisks* denote ions that were present in both the tandem mass spectroscopy and MS^3 spectra, which were due to y-type ions

the MS/MS spectrum of the peptide IPDGFAGAQGGITFR (Fig. 9.6, above) is compared with the MS^3 spectrum of the doubly charged y-type ion with an m/z of 697.3. Ions labelled with an *asterisk* appearing in both spectra have the same mass and are y-type ions. It is not unusual to find doubly charged y-type ions resulting from the loss of one or two N-terminal amino acids in an MS/MS spectrum obtained from an ion trap, and these are the best choice as precursors of the MS^3 spectrum. If doubly charged y-type ions are absent, then a singly charged y-type ion above the peptide precursor ion can be selected for further fragmentation. The difficulty with this approach is that, in an uninterpreted MS/MS spectrum, it is impossible to know if a y-type ion has been selected for the MS^3 spectrum. However, nanospray ionisation allows sufficient time for the acquisition of several MS^3 spectra, which increases the chance that a y-type ion will eventually be selected. MS^3 spectra of b-type ions may be problematic, as gas phase rearrangements may occur, given the longer reaction times in an ion trap.

9.5.6 Comparison with Synthetic Peptides

If the sample is of sufficient importance, it is also possible to make synthetic peptides of the proposed sequence. Fragmentation patterns of the unknown peptide can be compared with that of the synthetic peptide. Likewise, the high-performance liquid-chromatography retention times of the unknown and the synthetic peptide can be compared.

9.6 Conclusion

The aim of this chapter is to illustrate that the interpretation of MS/MS spectra is not simple. Although the mathematics involved requires no more than a third-grade education, there are enough obfuscating factors to make it impossible to determine a sequence with 100% confidence from a single MS/MS spectrum. However, partially correct sequences can be useful for verifying sequence-database matches and for performing homology-based database searches. If greater confidence is required in a sequence assignment (for cloning), there are a number of experiments that can be performed to reduce sequence ambiguity.

Acknowledgements

The author thanks Anne Aumell and Alex Taylor for their careful reading of the manuscript.

References

Altschul SF, Gish W, Miller W, Myers EW, Lipman DJ (1990) Basic local alignment search tool. J Mol Biol 215:403–410

Arnott D, Henzel W, Stults JT (1998) Rapid identification of comigrating gel-isolated proteins by ion trap mass spectrometry. Electrophoresis 19:968–980

Bahr U, Karas M, Kelner R (1998) Differentiation of lysine/glutamine in peptide sequence analysis by electrospray ionization sequential mass spectrometry coupled with a quadrupole ion trap. Rapid Commun Mass Spectrom 12:1382–1388

Bartels C (1990) Fast algorithm for peptide sequencing by mass spectroscopy. Biomed Environ Mass Spectrom 19:363–368

Biemann K, Cone C, Webster BR, Arsenault GP (1966) Determination of the amino acid sequence in oligopeptides by computer interpretation of their high-resolution mass spectra. J Am Chem Soc 88:5598–5606

Clauser KR, Baker P, Burlingame AL (1996) Peptide fragment ion tags from MALDI/PSD for error-tolerant searching of genomic databases. The 44th ASMS Conference on Mass Spectrometry and Allied Topics, Portland, OR

Davis MT, Lee TD (1997) Variable flow LC/MS/MS and the comprehensive analysis of complex protein mixtures. J Am Soc Mass Spectrom 8:1059–1069

Desiderio D, Kai M (1983) Preparation of stable isotope-incorporated peptide internal standards for field desorption mass spectrometry quantification of peptides in biologic tissue. Biomed Mass Spectrom 10:471–479

Eng JK, McCormack AL, Yates JR III (1994) An approach to correlate tandem mass spectra data of peptides with amino acid sequences in a protein database. J Am Soc Mass Spectrom 5:976–989

Fenyo D, Qin J, Chait BT (1998) Protein indentification using mass spectrometric information. Electrophoresis 19:998–1005

Fernandez-de-Cossio J, Gonzalez J, Besada V (1995) A computer program to aid the sequencing of peptides in collision-activated decomposition experiments. Comput Appl Biosci 11:427–434

Gaskell SJ, Haroldsen PE, Reilly MH (1988) Collisionally activated decomposition of modified peptides using a tandem hybrid instrument. Biomed Environ Mass Spectrom 16:31–33

Hamm CW, Wilson WE, Harvan DJ (1986) Peptide sequencing program. Comput Appl Biosci 2:115–118

Henzel WJ, Billeci TM, Stults JT, Wong SC, Grimley C, Watanabe C (1993) Identifying proteins from two-dimensional gels by molecular mass searching of peptide fragments in protein sequence databases. Proc Natl Acad Sci USA 90:5011–5015

Hines WM, Falick AM, Burlingame AL, Gibson BW (1992) Pattern-based algorithm for peptide sequencing from tandem high energy collision-induced dissociation mass spectra. J Am Soc Mass Spectrom 3:326–336

Hopper S, Johnson RS, Vath JE, Biemann K (1989) Glutaredoxin from rabbit bone marrow. Purification, characterization, and amino acid sequence determined by tandem mass spectrometry. J Biol Chem 264:20438–20447

Hulst AG, Kientz CE (1996) Differentiation between isomeric amino acids leucine and isoleucine using low energy collision-induced dissociation tandem mass spectrometry. J Mass Spectrom 31:1188–1190

Hunt DF, Yates JR III, Shabanowitz J, Winston S, Hauer CR (1986) Protein sequencing by tandem mass spectrometry. Proc Natl Acad Sci USA 83:6233–6237

Ishikawa K, Niwa Y (1986) Computer-aided peptide sequencing by fast atom bombardment mass spectrometry. Biomed Environ Mass Spectrom 13:373–380

Johnson RS, Biemann K (1987) The primary structure of thioredoxin from Chromatium vinosum determined by high-performance tandem mass spectrometry. Biochemistry 26:1209–1214

Johnson RS, Biemann K (1989) Computer program (SEQPEP) to aid in the interpretation of high-energy collision tandem mass spectra of peptides. Biomed Environ Mass Spectrom 18:945–957

Johnson RS, Martin SA, Biemann K, Stults JT, Watson JT (1987) Novel fragmentation process of peptides by collision-induced decomposition in a tandem mass spectrometer: differentiation of leucine and isoleucine. Anal Chem 59:2621–2625

Johnson RS, Mathews WR, Biemann K, Hopper S (1988a) Amino acid sequence of thioredoxin isolated from rabbit bone marrow determined by tandem mass spectrometry. J Biol Chem 263:9589–9597

Johnson RS, Ericsson L, Walsh KA (1991) LUTEFISK without the odor – a computer program for the interpretation of low energy CID spectra of electrosprayed peptide ions. The 39th ASMS Conference on Mass Spectrometry and Allied Topics, Nashville, TN

Johnson TC, Yee BC, Carlson DE, Buchanan BB, Johnson RS, Mathews WR, Biemann K (1988b) Thioredoxin from Rhodospirillum rubrum: primary structure and relation to thioredoxins from other photosynthetic bacteria. J Bacteriol 170:2406–2408

Mann M, Wilm M (1994) Error-tolerant identification of peptides in sequence databases by peptide sequence tags. Anal Chem 66:4390–4399

Morris HR, Paxton T, Dell A, Langhorne J, Berg M, Bordoli RS, Hoyes J, Bateman RH (1996) High sensitivity collisionally-activated decomposition tandem mass spectrometry on a novel quadrupole / orthogonal-acceleration time-of-flight mass spectrometer. Rapid Commun Mass Spectrom 10:889–896

Papayannopoulos IA (1995) The interpretation of collision-induced dissociation tandem mass spectra of peptides. Mass Spectron Rev 14:49–73

Pearson WR, Lipman DJ (1988) Improved tools for biological sequence analysis. Proc Natl Acad Sci USA 85:2444–2448

Rose K, Simona M, Offord RE, Prior CP, Otto B, Thatcher DR (1983) A new mass spectrometric C-terminal sequencing technique finds a similarity between gamma-interferon and alpha 2-interferon and identifies a proteolytically clipped gamma-interferon that retains full antiviral activity. Biochem J 215:273–277

Sakurai T, Matsuo T, Matsuda H, Katakuse I (1984) PAAS3: a computer program to determine probable sequence of peptides from mass spectrometric data. Biomed Mass Spectrom 11:396–399

Scarberry RE, Zhang Z, Knapp DR (1995) Peptide sequence determination from high-energy collision-induced dissociation spectra using artificial neural networks. J Am Soc Mass Spectrom 6:947–961

Schnolzer M, Jedrzejewski P, Lehmann WD (1996) Protease-catalyzed incorporation of 18O into peptide fragments and its application for protein sequencing by electrospray and matrix-assisted laser desorption/ionization mass spectrometry. Electrophoresis 17:945–953

Scoble HA, Biller JE, Biemann K (1987) A graphics display-oriented strategy for the amino acid sequencing of peptides by tandem mass spectrometry. Fresenius Z Anal Chem 327:239–245

Sepetov NF, Issakova OL, Lebl M, Swiderek K, Stahl DC, Lee TD (1993) The use of hydrogen-deuterium exchange to facilitate peptide sequencing by electrospray tandem mass spectrometry. Rapid Commun Mass Spectrom 7:58–62

Shevchenko A, Chernushevich I, Ens W, Standing KG, Thomson B, Wilm M, Mann M (1997) Rapid 'de novo' peptide sequencing by a combination of nanoelectrospray, isotopic labeling and a quadrupole/time-of-flight mass spectrometer. Rapid Commun Mass Spectrom 11:1015–1024

Siegel MM, Bauman N (1988) An efficient algorithm for sequencing peptides using fast atom bombardment mass spectral data. Biomed Environ Mass Spectrom 15:333–343

Takao T, Hori H, Okamoto K, Harada A, Kamachi M, Shimonishi Y (1991) Facile assignment of sequence ions of a peptide labelled with 18O at the carboxyl terminus. Rapid Commun Mass Spectrom 5:312–315

Tang X-J, Thibault P, Boyd RK (1993) Fragmentation reactions of multiply-protonated peptides and implications for sequencing by tandem mass spectrometry with low-energy collision-induced dissociation. Anal Chem 65:2824–2834

Taylor JA, Johnson RS (1997) Sequence database searches via de novo peptide sequencing by tandem mass spectrometry. Rapid Commun Mass Spectrom 11:1067–1075

Whaley B, Caprioli RM (1991) Identification of nearest neighbor peptides in protease digests by mass spectrometry for construction of sequence-ordered tryptic maps. Biol Mass Spectrom 20:210–214

Yates JR III, Griffin PR, Hood LE (1991) Computer aided interpretation of low energy MS/MS mass spectra of peptides. Academic Press, San Diego

10 Specific Detection and Analysis of Phosphorylated Peptides by Mass Spectrometry

MANFREDO QUADRONI

10.1 Introduction

For proteins, phosphorylation is the most important known post-translational modification that affects activity. Virtually any basic process of a eukaryotic cell is regulated at some point by the phosphorylation of one or more of its key protein components (Krebs 1994). The regulation of gene transcription, cell-cycle progression, cell division and proliferation, cell differentiation, cytoskeletal dynamics, energy storage and retrieval are all phosphorylation dependent. The functional relevance of protein phosphorylation is even more striking, considering its relatively small quantitative importance. Only a few percent of proteins are phosphorylated and, of these, only 0.03% are phosphorylated at a tyrosine residue.

The amino acids most often phosphorylated are serine, threonine and tyrosine, and the analysis of these events is the subject of the chapter. Phosphate transfer is also known to occur on histidine and aspartic-acid residues in the two-component signalling systems of prokaryotes and some eukaryotes (Wurgler-Murphy and Saito 1997). Phosphoaspartate and phosphoglutamate intermediates have been characterised as intermediates in energy-coupling mechanisms that drive, for example, membrane transporters, but these will not be considered in this chapter.

The state of phosphorylation of a protein is generally highly dynamic and is modulated antagonistically by protein kinases and protein phosphatases (Hunter 1995), whereby the activity of both classes of enzymes is subject to further complex regulatory phenomena. Protein kinases, the family of enzymes that catalyses the transfer of a phosphate group from a nucleotide to a target protein, constitute the largest family of proteins known, accounting for at least 2% of the genes in yeast (Hunter 1987; Hanks et al. 1988; Hunter and Plowman 1997). Their origins can probably be traced to the beginnings of life, because the mechanism and structural basis of their action are shared with the fundamental mechanisms of cellular energy storage.

At the molecular level, phosphorylation is known to act by turning on/off enzymatic activity (i.e. phosphorylase-B kinase, mitogen-assisted protein kinases, Src) or by re-directing the targeting of proteins to the various intracellular compartments (i.e. transcription factors). Phosphorylation also modulates protein–protein interactions and the formation of complexes (SH_2- and other phosphotyrosine-binding domain-containing proteins). Protein phos-phorylation/de-phosphorylation on Ser, Thr and Tyr usually occurs at residues that are exposed on the surface of the protein, often on loops or turns, ensuring accessibility for the kinase or phosphatase. The crystal structures of several enzymes have been solved in the phospho and non-phospho forms, showing that the addition of a phosphate group can act in several possible ways to directly affect enzyme activity (Johnson and O'Reilly 1996; Johnson et al. 1998). Alternatively, especially in the case of tyrosine phosphorylation, the addition of the phosphate group at a certain position creates a "molecular tag". This is usually a binding site for a regulatory molecule that can inhibit or activate the phosphorylated partner or target it to a different intracellular compartment or a multimeric complex.

10.1.1 Experimental Approaches to the Study of Phosphorylation Sites in Proteins

Indications that a polypeptide is phosphorylated are generally obtained by in vivo or in vitro labelling with radioactive phosphate (^{32}P or ^{33}P) or by using monoclonal antibodies raised against generic phosphoamino acids (especially anti-phosphotyrosine) or specific amino acid sequences in the phosphorylated form. The most common problem facing the researcher is the determination of the number and location of the phosphates in the sequence. The degree of difficulty of this task depends on three main factors: (1) the amount and purity of the protein of interest, (2) the stoichiometry and distribution of phosphoamino acids and (3) the sequence near the phosphorylation site(s).

After obtaining the sample, the next step is usually specific proteolysis by enzymatic or chemical methods followed by separation of the resulting peptides and isolation of the phosphorylated ones. These are then subjected to sequence analysis. Classical methods rely on the separation of the proteolytic mixture by high-performance liquid chromatography (HPLC) or a combination of thin-layer electrophoresis and chromatography, and all require radioactive labelling with ^{32}P to identify the phosphorylated peptides. These techniques and those employing Edman degradation to locate the phosphorylated residue have been described extensively elsewhere (Meyer et al. 1986; Van der Geer and Hunter 1994; Fischer et al. 1997) and will not be discussed here. Instead, we will concentrate on the application of biological mass spectrometry (MS) to the analysis of phosphorylation sites in proteins. When compared with classical methods, MS has some decisive advantages, such as increased speed and sensitivity, the ability to obtain a larger amount of information, and the ability to avoid radioactive labelling.

10.2 Specific Detection of Phosphopeptides in Complex Peptide Mixtures

Even when a protein is phosphorylated in vitro with a highly active protein kinase, the reaction rarely proceeds to completion. In many cases, the final fraction of phosphorylated molecules is lower than 10%, and the percentage is usually even lower when the material stems from in vivo phosphorylation (heavily phosphorylated proteins, such as casein and profilaggrin, are a notable exception). The only cases for which it is possible to start with a fully phosphorylated sample are where the phosphoprotein has been purified by iso-electric focusing or 2D electrophoresis or by immunoaffinity-based isolation with an antibody specific for the phosphorylated form of the protein. Thus, finding minute amounts of phosphorylated peptides in a mixture of the dozens or hundreds of peptides present after digestion is sometimes a "needle in a haystack" problem.

10.2.1 Isolation of Phosphopeptides Prior to MS Analysis

The only general technique developed to isolate or enrich phosphorylated peptides from the mixture is immobilised-metal affinity chromatography (IMAC; Andersson and Porath 1986), which exploits the affinity of phosphoamino-acid side chains for chelated Fe^{3+}. This approach is best performed as described by Watts et al. (1994), using a micro-column to ensure maximum recovery when starting with small amounts of material. The Appendix (Sect. 1) outlines the steps and reagents required for performing separations on an IMAC column. It has to be noted, however, that the binding of phosphopeptides to the chelated Fe^{3+} column is strongly pH dependent and can be greatly influenced by the presence of competing buffers and salt molecules. Thus, the composition of the sample to be loaded is critical for the success of the purification. In addition, strongly acidic peptides containing clusters of aspartate or glutamate residues exhibit a certain degree of binding to IMAC columns, resulting in a variable degree of contamination by non-phosphorylated peptides.

The use of a new type of chelating resin based on nitrilotriacetic instead of iminodiacetic groups has helped to reduce contamination by non-phosphorylated peptides (Neville et al. 1997). One important aspect to consider when dealing with phosphopeptides is their relatively poor retention on reversed-phase (RP) resins. Short and very hydrophilic phosphopeptides in particular bind poorly to C_8 and POROS beads and columns. Care has to be taken to prevent losses during HPLC analysis or while de-salting samples on micro-columns prior to MS analysis. This problem has been analysed in detail by Neubauer and Mann (1999), who showed that the use of the more hydrophobic OligoR3 instead of R2-POROS resin greatly improves the recovery of phosphopeptides, although special conditions have to be used for elution. When dealing with very small amounts of sample, care should be taken to minimise sample contact with metallic components, such as syringe needles and HPLC tubing, to which non-specific absorption may occur due to the affinity of phosphopeptides for iron.

As an alternative to preparative methods, a method of detection that can specifically pinpoint the phosphorylated peptides after or even before the mixture has been separated into its individual components is needed. MS can be used to determine the composition of a complex peptide mixture; the identity of a peptide of a known mass can be defined if the sequence of the protein from which it is derived is known, together with the specificity of the protease used. In addition, in a triple-quadrupole, ion-trap or quadrupole time-of-flight (TOF) mass spectrometer, individual peptides can be isolated and subjected to tandem MS (MS/MS) measurements, which can provide information on the sequence and eventual modifications of the peptide backbone. Still, even with MS, it is often difficult to identify phosphorylated peptides after a simple analysis of the total composition of a digest. More sophisticated MS techniques are required to unequivocally identify phosphorylated peptides, especially when these are present in minute amounts.

10.2.2 Enzymatic Removal of the Phosphate Group

One straightforward technique for the identification of phosphopeptides is based on matrix-assisted laser desorption ionisation (MALDI)-TOF analysis of the protein digest before and after dephosphorylation by alkaline phosphatase (Liao et al. 1994). Peptides whose mass decreases by a multiple of 80 amu after the enzymatic treatment are phosphorylated. The procedure is very simple and fast and yields easily interpretable data. However, this method requires that the phosphopeptide be clearly detectable as a resolved peak by MALDI. In practice, because of the low stoichiometry of phosphorylation, phosphopeptides are often present in minute amounts (with signal levels near or below background noise) relative to the rest of the proteolytic fragments. In addition, suppression effects in MALDI-MS are common (Kratzer et al. 1998), and phosphopeptides tend to yield poor signals in positive-mode MALDI-TOF, although this can be partially compensated for by the presence of ammonium ions (Asara and Allison 1999). In conclusion, this approach is probably most suitable when at least microgram amounts of proteins are available for phosphorylation in vitro at relatively high stoichiometry.

10.2.3 Phosphopeptide Detection by Selective MS Scanning Methods

Techniques to specifically detect phosphopeptides in complex mixtures have been developed quite recently and exploit the ability of triple-stage quadrupole mass spectrometers to perform so-called precursor-ion (parent) scan and neutral-loss measurements. In a parent scan, the third quadrupole mass filter (Q3) is set to transmit only ions with a specific mass/charge ratio (m/z). At the same time, the first quadrupole (Q1) scans the whole mass range, allowing all ions present in the sample to enter the instrument in succession according to their m/z. These ions are fragmented in the second quadrupole (Q2), which is filled with inert gas and acts as a collision chamber (Chap. 2). The first and third quadrupoles are linked to one another and to the detector, so the latter records which ions entering the instrument produce the ion of interest after fragmentation in Q2.

In a neutral-loss measurement, quadrupoles Q1 and Q3 are scanned with a constant mass difference M corresponding to the mass of a group that fragments during collision-induced dissociation (CID) in Q2. Thus, a neutral-loss scan has the ability to detect the loss of an uncharged fragment of the desired mass M, while parent scanning is used to detect parent ions that generate a charged fragment on CID. To define a way to detect phosphorylated peptides, it is necessary to concentrate on the properties of the phosphoamino-acid side chains, especially their behaviour under CID conditions. The side chains of phosphoserine, phosphothreonine or phosphotyrosine can fragment easily on both sides of the phosphoester bond after collision with gas molecules. The masses of the fragments that we can expect to be produced, depending whether the sample is injected under acidic or basic conditions, are shown in Fig. 10.1.

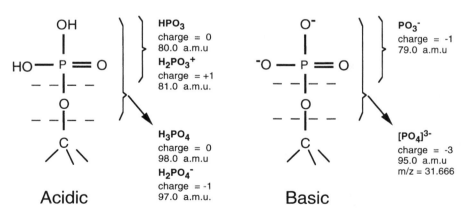

Fig. 10.1. Collision-induced dissociation (CID) fragments of the phosphoamino-acid side chain under acidic or basic conditions. The potential fragments produced by a phosphorylated amino acid side chain subjected to CID or post-source decay are shown. Acidic conditions are used with positive-mode mass spectrometry (MS); basic conditions are used with negative-mode MS. In practice, only neutral loss of H_3PO_4 ($\Delta m = -98$ amu) is observed under acidic conditions (positive mode), while the PO_3^- ion (m/z − 98 amu) is efficiently produced at high pH and is detected in negative-mode MS

The charged fragments can be detected in a parent-scan measurement, while the uncharged ones can be inferred from neutral-loss measurements. In practice, under CID conditions, some of these possible fragments are not produced in sufficient amounts to be useful (as is the case for $[H_2PO]^{3+}$). Others are of no practical utility, because they coincide with fragment ions produced abundantly by other amino acid side chains (97 and 99 amu correspond to b1 ions from Pro and Val, respectively). Neutral-loss monitoring for a Δm of 98 amu has been applied occasionally (Huddleston et al. 1993) but generally suffers from low sensitivity and specificity.

Parent scanning for the production of m/z = 79 in negative mode (Carr et al. 1996; Wilm et al. 1996) has proven to be the best method for the specific detection of native phosphorylated peptides in complex mixtures. It combines high selectivity with good sensitivity. The problem for most users is the necessity of measuring in negative-ion mode, a fairly unusual condition that is not implemented in most protein-analysis laboratories and is not easily made compatible with on-line separation methods based on RP liquid chromatography. Therefore, it has been applied exclusively in conjunction with an off-line nanospray source that allows the pH of the sample to be switched easily from acidic to basic and vice versa. The essential steps for the identification of phosphorylated peptides according to this procedure are described in Fig. 10.2, which also shows the accumulation of a parent scan with an m/z of 79 in negative mode at a pH greater than 8.0. After data accumulation, the sample is acidified, and the peptides that produced a parent-scan signal are selected for MS/MS analysis in positive-ion mode. In this regard, it is worth emphasising that the mass spectrometer must be suitable for high-sensitivity parent-scanning measurements, and it is necessary to specifically optimise a tuning table for parent scanning in negative mode in order to obtain stable and intense signals. A triple-quadrupole operation in parent-ion mode is intrinsically much more sensitive that one operating in scanning mode, because the detector continuously monitors a single ion and does not scan. Thus, signals from phosphopeptides are often seen in parent-ion mode that cannot be seen in normal scanning mode.

10.2.4 Phosphopeptide Detection by Selective MS Scanning Methods After Specific Chemical Modifications

An alternative to analysis in negative mode is the chemical derivatisation of phosphorylated side chains in order to substitute them with a group that can act as a "mass marker". We have developed a technique to convert phosphoserine and phosphothreonine into S-pyridylethylcysteine and S-pyridylethyl-β-methylcysteine, respectively. The thioether bond breaks easily under CID conditions, producing the pyridylethyl ion (m/z = 106) in a highly efficient manner. The first step in the derivatisation is the $Ba(OH)_2$-catalysed β-elimination of the phosphate group (Holmes 1987; Byford 1991; Resing et al.

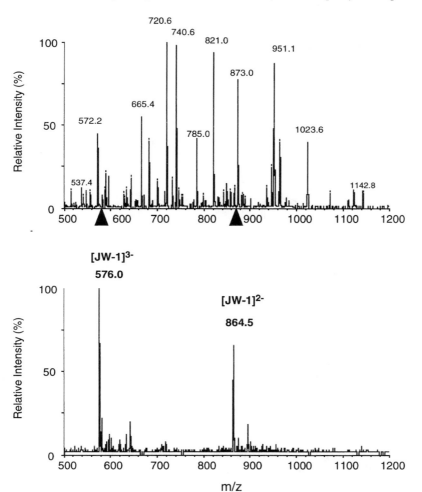

Fig. 10.2. Selectivity of parent-ion scanning for the production of the fragment PO_3^-. The synthetic phosphopeptide Jw-1 [GGGSLKAPSRPAIP(pS)LT; mass (average) = 1730.87 amu, final concentration = 0.2 pmol/µl] was mixed with a tryptic digest of bovine serum albumin (5 pmol/µl), and 1 µl of the resulting mixture (50% methanol, 1% acetic acid; pH 4) was analysed by a full scan in positive mode (*above*) by nanospray mass spectrometry (Wilm and Mann 1996) on a SCIEX API-III+ triple-quadrupole mass spectrometer. The sample was then acidified by the addition of 0.2 µl of 5% ammonia and was then analysed in negative mode to acquire the parent scan with an m/z of 79 (*below*). The peaks for the 2+ and 3+ charge states of the phosphopeptide are barely visible in positive mode (*triangles*) but dominate the parent-scan spectrum

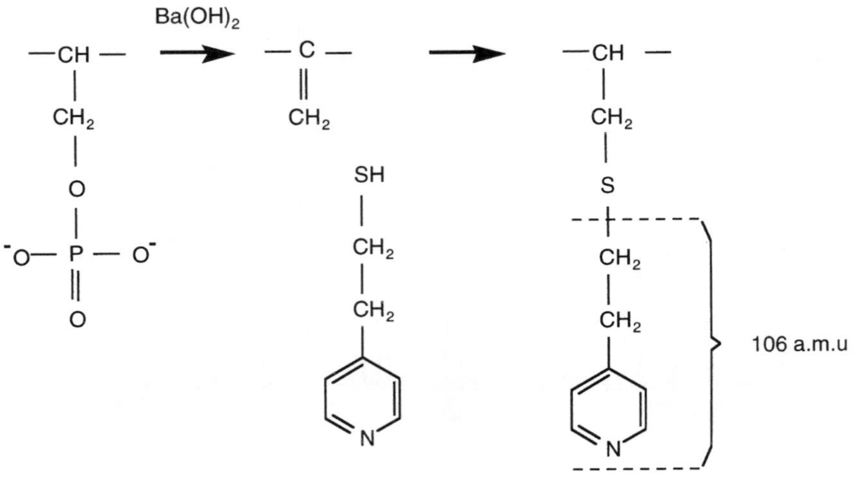

Fig. 10.3. Modification of phosphoserine to introduce a mass tag. Phosphoserine is converted into *S*-pyridylethylcysteine upon Ba(OH)$_2$-catalysed β-elimination and reaction with 2-[4-pyridyl]ethanethiol. The reaction is performed in one step, as described in the Appendix (Sect. 2), where the synthesis of 2-[4-pyridyl]ethanethiol is also described

1995), which produces dehydroalanine or dehydrothreonine. The former reacts quickly and completely with 2-[4-pyrydyl]-ethanethiol to yield the required derivatives (Fig. 10.3).

The reaction proceeds significantly slower in the case of dehydrothreonine, but we found that a reaction time of 3 h is sufficient to obtain quantitative conversion to the *S*-pyridylethylated side chain (Appendix, Sect. 2). An example of this specific modification is shown in Fig. 10.4 using the model protein β-casein. Mature β-casein is phosphorylated on five serines, at positions 15, 17, 18, 19 and 35. After digestion with trypsin, however, only two phosphopeptides are generated, namely the tetraphosphorylated P1 [residues 1–25, RELEEL-NVPGEIVE(pS)L(pS)(pS)(pS)EESITR; MH$^+$(average) = 3122.95 amu] and the mono-phosphorylated P2 [residues 33–48, FQ(pS)EEQQQTEDELQDK; MH$^+$(average) = 2061.98 amu]. Figure 10.4 shows a full scan and a parent scan for the fragment ion with an m/z of 106 for a tryptic digest of β-casein after full conversion of the phosphoserines to *S*-pyridylethylcysteines (Appendix, Sect. 2). The modification results in a change in mass of +41 amu per site, so the calculated masses of the fully derivatised P1 and P2 (named P1* and P2*) are 3286.95 and 2102.98 amu, respectively. In the parent (m/z = 106) scan, we only observe signals corresponding to the 3+ charge state of P1* and the +2 state of P2*. The advantages of the conversion to *S*-pyridylethylcysteine include the specific detectability by parent scanning, the ability to constantly work in positive mode and an increase in signal intensity in positive-ion mode, due to the replacement of the negatively charged or neutral phosphate

A)

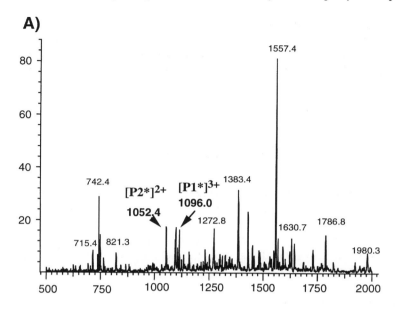

B)

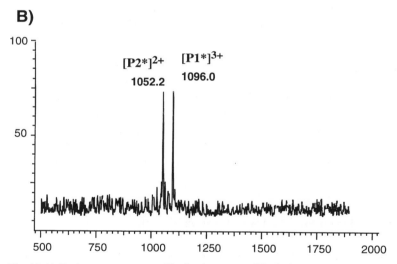

Fig. 10.4A,B. A parent scan specifically detects modified phosphopeptides. A tryptic digest of bovine β-casein is analysed by a full scan (**A**) and a parent scan for the production of the fragment with an m/z of 106 (**B**) after the conversion of phosphoserine residues to S-pyridylethyl-cysteine, as described in the Appendix (Sect. 2). The 2+ and 3+ charge states of modified peptides P2 and P1 (*P2** and *P1**) are clearly detected

group with the easily protonatable S-pyridylethyl chain. The main disadvantages of this method are the need to perform a complex chemical modification with the potential for side reactions and, obviously, the fact that it is not feasible for phosphotyrosine. In addition, previous reduction and alkylation of cysteines is theoretically required to prevent intramolecular cross-linking.

10.3 Phosphorylation-Site Analysis by MS/MS

By phosphorylation-site analysis, we mean the detailed study of phosphorylated peptides to determine the type of phosphorylation (Ser, Thr or Tyr) and the precise location of the phosphoamino acid(s) in the sequence. Again, we will concentrate on methods based on MS, because more classical methods using radioactive labelling and Edman degradation have been described in detail elsewhere (Van der Geer and Hunter 1994; Fischer et al. 1997).

In general, analysis of the phosphorylation site(s) by MS involves the fragmentation of the phosphopeptide, followed by the measurement of the masses of the resulting fragments. Two techniques are mainly used for this purpose: MS/MS on a triple-stage quadrupole or ion-trap MS and PSD (post-source decay) on a MALDI-TOF instrument. In general, MS/MS appears to be a more generally applicable technique than PSD and usually produces clearer spectra with a higher information content. Although the techniques for performing PSD are evolving rapidly, the production of good PSD spectra seems to be strongly sequence dependent and is strongly influenced by sample purity. The fraction of peptides in a digest that produce useful spectra is usually very low, though chemical modifications are being developed to overcome this (Chap. 3). However, PSD can be successfully used to discriminate between Ser/Thr and Tyr phosphorylation (Annan and Carr 1996), as we will discuss in the next section.

10.3.1 Phosphorylation-Site Analysis of Native Phosphopeptides

The masses of phosphoserine, phosphothreonine and phosphotyrosine are, respectively, 167, 181 and 243 amu; thus, they do not coincide with those of any other naturally occurring amino acid. Thus, in principle, a series of b- or y-type ions (Chap. 8) covering the sequence around the phosphorylatable residue(s) is enough to exactly pinpoint the modification.

For peptides containing a single phosphoamino acid, it is possible to determine the type (Ser/Thr or Tyr) of phosphorylation via a close analysis of their PSD spectra; Annan and Carr (1996) reported that PSer and PThr peptides tend to produce (by decay) an abundant $[MH-H_3PO^4]^+$ peak (MH$^+$−98 amu) and only a small [MH–HPO^3] ion (MH$^+$–80 amu), whereas the ratio is inverted for phosphotyrosine peptides. This is easily understandable in terms of the structures of the side chains.

Phosphopeptides, especially when multiply phosphorylated, are poorly protonatable and, upon CID, often produce MS/MS spectra that are either dominated by the molecular ion minus phosphate(s) or are complicated by the partial neutral loss of the phosphate group(s), which generates multiple ion series. The extent of this last phenomenon is highly variable and depends primarily on the sequence, the collision energy and the type and pressure of the collision gas. Two diametrically opposed examples (a "regular" spectrum and one strongly affected by the neutral loss of phosphate) are shown in Fig. 10.5. The spectrum of the P2 peptide from β-casein is characterised by series of b- and y-type ions at the expected masses (Fig. 10.5A), together with b* and y* ions resulting from the neutral loss of phosphate. Instead, the MS/MS of the synthetic peptide ASGQAFELIL(pS)PR, corresponding to residues 14–27 of the phosphoprotein OP18/stathmin, has an almost complete series of dominating (y–98) ions, resulting from extensive neutral loss of the phosphate group.

As mentioned earlier, this phenomenon becomes a problem when dealing with multiple phosphorylated peptides. Ways of addressing it employ mostly chemical modification of the phosphoamino acids and are discussed in the next section. The only other possibility is to exploit the ability of ion-trap instruments to perform multiple consecutive stages of fragmentation (i.e. MS/MS/MS or MS^3). This would allow one to isolate "in vacuo" de-phosphorylated forms of the parent ion after a first round of CID and to subject them to a further stage of fragmentation that would allow the measurement of "normal" y- and b-type ion series.

10.3.2 Phosphorylation-Site Analysis After Chemical Modification of the Phosphoamino-Acid Side Chains

The general goals of methods that modify the phosphorylated side chains are: (1) to increase the detectability of phosphopeptides in positive-ionisation mode, (2) to incorporate a "mass tag" that enables the specific detection of the phosphopeptides through parent scanning and (3) to simplify product-ion spectra to facilitate their interpretation.

Methods of specifically modifying phosphoamino acids usually target phosphoserine and phosphothreonine and exploit their tendency to undergo β-elimination at a very basic pH to yield dehydroalanine and dehydrothreonine, resulting in a decrease in mass of 98 amu. The reaction proceeds much faster when catalysed by Ba^{2+} ions (Byford 1991), so a 2- to 3-h incubation at 30 °C is sufficient to induce the quantitative elimination of phosphate groups. This method has been applied (Resing et al. 1995) to the study of profilaggrin, a protein containing up to 400 mol phosphate/mol protein. Resing et al. found

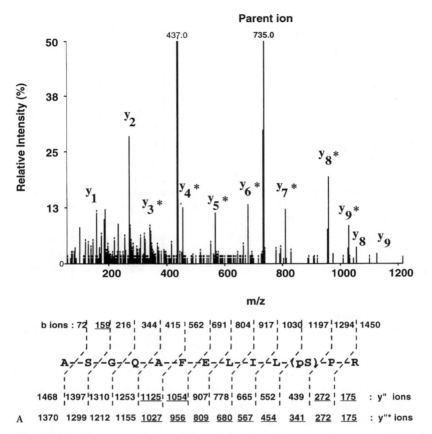

Fig. 10.5A,B. The extent of neutral loss of the phosphate group is sequence dependent. The collision-induced spectrum of the synthetic peptide ASGQAFELIL(pS)PR (**A**) shows that this peptide has a strong tendency to lose the whole phosphoric acid group (H₃PO₄; Δm = –98 amu), thereby generating an almost complete series of y"–98 (y*) ions. By contrast, peptide P2 from β-casein [FQ(pS)EEQQQTEDELQDK] displays a tandem mass-spectrometry fragmentation pattern (**B**) that is only partly affected by phosphate loss and shows most y" and b-type ions predicted from the sequence. Ions detected in the spectra are *underlined*

that the modification improves the chromatographic behaviour of peptides, increases the recovery of short hydrophilic phosphopeptides and greatly improves the quality of MS/MS spectra. One caveat is the instability of peptides containing dehydroalanine and dehydrothreonine, because these are very reactive and tend to cross-link and adsorb to any material, so they are best analysed as soon as possible.

A variation of the β-elimination method is subsequent derivatisation with 2-[4-pyridyl]ethanethiol, which has the advantage of adding a mass tag for the specific identification of modified peptides by parent scanning. The procedure

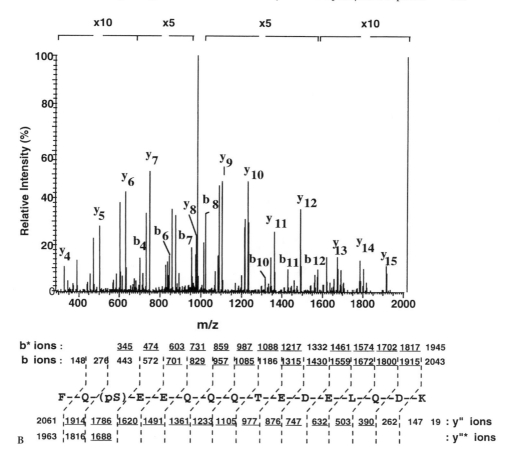

b* ions :   345   474   603 731   859   987 1088 1217 1332 1461 1574 1702 1817 1945
b ions : 148 276 443 572 701 829 957 1085 1186 1315 1430 1559 1672 1800 1915 2043

F–Q–(pS)–E–E–Q–Q–Q–T–E–D–E–L–Q–D–K

2061 1914 1786 1620 1491 1361 1233 1105 977 876 747 632 503 390 262 147 19 : y" ions
B 1963 1816 1688   : y"* ions

Fig. 10.5. *Continued*

for such an experiment is reported above and in the Appendix (Sect. 2). Figure 10.6 shows that modification of the tetraphosphorylated P1 peptide from β-casein results in a large (~100-fold) increase in the signal generated by this peptide in electrospray-ionisation MS and thus allows the acquisition of a good MS/MS spectrum, which is impossible to obtain from the native peptide.

10.4 Data Analysis

Recent developments in software for MS data analysis have resulted in computational tools that can help in the interpretation of MS/MS spectra of phosphopeptides and in their assignment to a specific amino acid

A ms/ms 1042.0

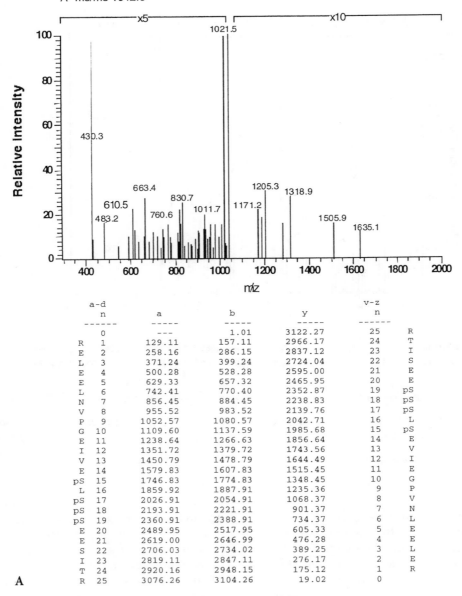

a–d n	a	b	y	v–z n	
0	---	1.01	3122.27	25	R
R 1	129.11	157.11	2966.17	24	T
E 2	258.16	286.15	2837.12	23	I
L 3	371.24	399.24	2724.04	22	S
E 4	500.28	528.28	2595.00	21	E
E 5	629.33	657.32	2465.95	20	E
L 6	742.41	770.40	2352.87	19	pS
N 7	856.45	884.45	2238.83	18	pS
V 8	955.52	983.52	2139.76	17	pS
P 9	1052.57	1080.57	2042.71	16	L
G 10	1109.60	1137.59	1985.68	15	pS
E 11	1238.64	1266.63	1856.64	14	E
I 12	1351.72	1379.72	1743.56	13	V
V 13	1450.79	1478.79	1644.49	12	I
E 14	1579.83	1607.83	1515.45	11	E
pS 15	1746.83	1774.83	1348.45	10	G
L 16	1859.92	1887.91	1235.36	9	P
pS 17	2026.91	2054.91	1068.37	8	V
pS 18	2193.91	2221.91	901.37	7	N
pS 19	2360.91	2388.91	734.37	6	L
E 20	2489.95	2517.95	605.33	5	E
E 21	2619.00	2646.99	476.28	4	E
S 22	2706.03	2734.02	389.25	3	L
I 23	2819.11	2847.11	276.17	2	E
T 24	2920.16	2948.15	175.12	1	R
R 25	3076.26	3104.26	19.02	0	

A

Fig. 10.6A,B. Conversion of phosphoserine to S-pyridylethylcysteine greatly enhances the signal intensity for a multiply phosphorylated peptide in electrospray-ionisation mass spectrometry (ESI-MS). Phosphopeptides P1 and P2 were purified from a tryptic digest of bovine β-casein by immobilised-metal affinity chromatography (Appendix, Sect. 1), desalted on a reversed-phase column and analysed by ESI-MS. The tetraphosphorylated P1 peptide [RELEEL-NVPGEIVE(pS)L(pS)(pS)(pS)EESITR] was very poorly detected, even at a very low pH (0.5% trifluoroacetic acid, pH < 2), and tandem MS analysis produced a spectrum with a very low information content (**A**). Interestingly, P1 in the same sample was clearly detectable using matrix-assisted laser desorption ionisation time-of-flight MS. The same fraction was then subjected to β-elimination and the addition of 2-[4-pyridyl] ethanethiol and was re-analysed as in **A**. The modified P1 (P1*; MH⁺ = 3286.95 amu), carrying four S-pyridylethylcysteine residues, was readily detectable by ESI-MS and produced a highly informative spectrum (**B**). Ions found in the spectrum are labelled with an asterisk

B ms/ms 1097.0 (3+)

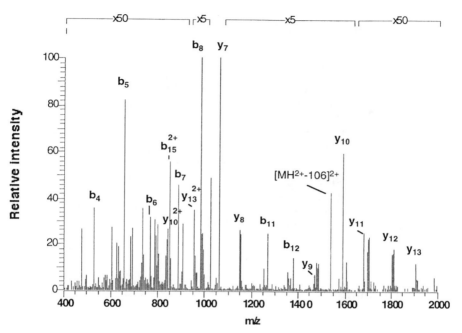

X = S-pyrydylethylcysteine

	a–d n	a	b	y	v–z n	
	0	---	1.01	3286.54	25	R
R	1	129.11	157.11	3130.44	24	T
E	2	258.16	286.15	3001.40	23	I
L	3	371.24	399.24	2888.31	22	S
E	4	500.28	528.28*	2759.27	21	E
E	5	629.33	657.32*	2630.23	20	E
L	6	742.41	770.40*	2517.14	19	X
N	7	856.45	884.45*	2403.10	18	X
V	8	955.52*	983.52*	2304.03	17	X
P	9	1052.57	1080.57	2206.98	16	L
G	10	1109.60	1137.59	2149.96	15	X
E	11	1238.64*	1266.63*	2020.92	14	E
I	12	1351.72	1379.72*	1907.83*	13	V
V	13	1450.79	1478.79*	1808.76*	12	I
E	14	1579.83	1607.83*	1679.72*	11	E
X	15	1787.90	1815.90*	1471.65*	10	G
L	16	1900.98	1928.98	1358.57*	9	P
X	17	2109.05	2137.05	1150.50*	8	V
X	18	2317.12	2345.11	942.44*	7	N
X	19	2525.19	2553.18	734.37*	6	L
E	20	2654.23	2682.22	605.33*	5	E
E	21	2783.27	2811.27	476.28*	4	E
S	22	2870.30	2898.30	389.25	3	L
I	23	2983.39	3011.38	276.17	2	E
T	24	3084.43	3112.43	175.12	1	R
R	25	3240.53	3268.53	19.02	0	

B

Fig. 10.6. *Continued*

sequence. The database-searching program SEQUEST (Yates et al. 1995; http://thompson.mbt.washington.edu/sequest) matches raw MS/MS spectra to protein sequences using the most common fragment ions. One interesting option of the program is the ability to take into account "differential modifications", i.e. changes in the molecular weights of amino acids (+80.0 for phosphorylation) that may affect some (but not necessarily all) residues of a given type (i.e. Ser, Thr and Tyr). With this approach, it is theoretically possible, starting with a phosphopeptide MS/MS spectrum, to identify the protein from which the peptide originated, the exact sequence and the phosphorylation site.

The program SHERPA (Taylor et al. 1996; http://www.lsbc.com:70) can also perform similar matching calculations and, although it is not suitable for screening large sequence databases, it can perform more detailed comparisons, taking into account, for example, internal ions. We found it to be the best choice for mapping phosphorylation sites in a known sequence. The help provided by these and other software tools allows one to speed the data-analysis procedure considerably. These programs (and similar ones) will become indispensable in the near future.

10.5 Conclusion

The completion of the human genome sequence will bring a considerable advance in our knowledge of the whole repertoire of eukaryotic protein kinases. Extensive consensus-sequence-based searches of potential substrates for a given kinase can be performed once the whole genome is known. Still, there will be a great demand to test (in vitro or in vivo) whether such predicted phosphorylation events actually take place, in addition to when and under what conditions the phosphorylation occurs. In fact, the need to perform phosphorylation-site studies may actually increase to such an extent that an advanced degree of automation will be required to allow high-throughput analysis of peptide or protein libraries.

It is likely that analysis of phosphorylation sites in vivo will coincide with advances in proteome analysis by creating extensive databases of ^{32}P-labelled proteins as they are resolved by two-dimensional polyacrylamide-gel electrophoresis. Some studies applying a similar approach to signal-transduction research have already been published (Guy et al. 1994; Soskic et al. 1999). We predict that MS and some of the techniques we outlined in this chapter will be the basis of investigations during the post-genome era of phosphoprotein analysis.

Appendix

1 IMAC Chromatography for the Isolation of Phosphorylated Peptides

1.1 Column Packing

1. Micro-columns can be packed (as described in Watts et al. 1994) into 0.0625 in (1.6 mm) ×0.01 in (250 μm) Teflon tubing closed at the ends by the insertion of 280 × 50 μm fused-silica capillary tubing that is then glued in place with epoxy resin. The micro-column can be connected to an HPLC pump via standard fittings.
2. Mini-columns can be packed in plastic pipette tips (100 or 1000 μl) by the insertion of a fibreglass filter or glass wool at the tip. Packed columns offer a certain resistance to flow and suction may prove necessary to ensure a good flow.

1.2 IMAC Separation

Stock solutions:

- 30 mM $FeCl_3$ in high-purity H_2O
- 0.1 M Acetic acid
- Buffer A: 0.1% ammonium acetate (pH 8.0), adjusted with ammonia
- Buffer B: 0.1% ammonium acetate (pH 8) and 500 mM Na_2HPO_4 (pH 8.5)
- Buffer C: elution buffer [9:1 (v/v) A:B (v/v)]

1. Equilibrate the Chelating Sepharose Fast Flow (Pharmacia) in a fivefold excess of water
2. Pack the resin in a micro-column, as described above
3. Flush the column with 20 volumes of high-purity H_2O
4. Saturate the column with $FeCl_3$; inject five portions of 30-mM $FeCl_3$, one bed volume each
5. Flush extensively with H_2O to remove excess $FeCl_3$
6. Equilibrate with 0.1-M acetic acid (minimum ten bed volumes)
7. Load the peptide mixture (dissolved in 0.1-M acetic acid, possibly with a low salt and buffer content)
8. Flush with 0.1-M acetic acid (ten bed volumes)
9. Elute with three bed volumes of elution buffer

After elution, the fraction can be dried, re-suspended in 0.1% trifluoroacetic acid and analysed by RP-HPLC or de-salted for direct MS analysis. De-salting is not necessary for MALDI-TOF analysis. Phosphopeptides bind to the Fe^{3+}-loaded IMAC at a pH of 3 and elute at a pH of 8–9.

2 Chemical Modification of Phosphopeptides for MS

2.1 Synthesis of 2-[4-pyridyl]ethanethiol

1. Saturate 100 ml pyridine with H_2S by bubbling the gas produced by a Kipp apparatus through the solution for 15 min
2. Slowly add 4.0 ml of 4-vinylpyridine with stirring, with H_2S continuously bubbling through the reaction mixture
3. Saturate the mixture with H_2S for at least two more hours
4. Eliminate the pyridine by rotary evaporation, leaving approximately 4.5 ml of yellow, dense liquid composed of approximately 95% 2-[4-pyridyl]ethanethiol and 5% pyridine
5. Store the 2-[4-pyridyl]ethanethiol at 4 °C in an argon atmosphere

2.2 Conversion of Phosphoserine to S-Pyrydylethylcysteine

1. Reduce and carboxymethylate the cysteines in the protein or peptide
2. Re-suspend the lyophilised sample in saturated $Ba(OH)_2$ in high-purity H_2O
3. Add 2-[4-pyridyl]ethanethiol to a final concentration of 0.5% (v/v)
4. Vortex the turbid mixture and place in a bath sonicator for 1 min
5. Overlay with argon and incubate for 90 min at 40 °C
6. Add 5% acetic acid, after which the solution turns clear
7. Dilute the sample with three volumes of 0.1% acetic acid and inject onto an HPLC column or de-salt on a C_{18} or POROS micro-column

De-salting is necessary for MALDI-TOF analysis, because the presence of $Ba(OH)_2$ causes numerous artefactual peaks at masses greater than 1000 Da.

References

Andersson L, Porath J (1986) Isolation of phosphoproteins by immobilized metal (Fe^{3+}) affinity chromatography. Anal Biochem 154:250–254

Annan RS, Carr SA (1996) Phosphopeptide analysis by matrix-assisted laser desorption time-of-flight mass spectrometry. Anal Chem 68:3413–3421

Asara JM, Allison J (1999) Enhanced detection of phosphopeptides in matrix-assisted laser desorption/ionization mass spectrometry using ammonium salts. J Am Soc Mass Spectrom 10:35–44

Byford MF (1991) Rapid and selective modification of phosphoserine residues catalysed by Ba2+ ions for their detection during peptide microsequencing. Biochem J 280:261–265

Carr SA, Huddleston MJ, Annan RS (1996) Selective detection and sequencing of phosphopeptides at the femtomole level by mass spectrometry. Anal Biochem 239:180–912

Fischer WH, Hoeger CA, Meisenhelde, J, Hunter T, Craig AG (1997) Determination of phosphorylation sites in peptides and proteins employing a volatile Edman reagent. J Prot Chem 16:329–334

Guy GR, Philip R, Tan YH (1994) Analysis of cellular phosphoproteins by two-dimensional gel electrophoresis: applications for cell signaling in normal and cancer cells. Electrophoresis 15:417–440

Hanks SK, Quinn AM, Hunter T (1988) The protein kinase family: conserved features and deduced phylogeny of the catalytic domains. Science 241:42–52

Holmes CF (1987) A new method for the selective isolation of phosphoserine-containing peptides. FEBS Lett 215:21–24

Huddleston MJ, Annan RS, Bean MF, Carr SA (1993) Selective detection of phosphopeptides in complex mixtures by electrospray liquid chromatography/mass spectrometry. J Am Soc Mass Spectrom 4:710–717

Hunter T (1987) A thousand and one protein kinases. Cell 50:823–829

Hunter T (1995) Protein kinases and phosphatases: the yin and yang of protein phosphorylation and signaling. Cell 80:225–236

Hunter T, Plowman GD (1997) The protein kinases of budding yeast: six score and more. Trends Biochem Sci 22:18–22

Johnson LN, O'Reilly M (1996) Control by phosphorylation. Curr Opin Struct Biol 6:762–769

Johnson LN, Lowe ED, Noble MEN, Owen D (1998) The Eleventh Datta Lecture: the structural basis for substrate recognition and control by protein kinases. FEBS Lett 430:1–11

Kratzer R, Eckerskorn C, Karas M, Lottspeich F (1998) Suppression effects in enzymatic peptide ladder sequencing using ultraviolet – matrix assisted laser desorption/ionization – mass spectrometry. Electrophoresis 19:1910–1919

Krebs EG (1994) The growth of research on protein phosphorylation. Trends Biochem Sci 19:439

Liao PC, Leykam J, Andrews PC, Gage DA, Allison J (1994) An approach to locate phosphorylation sites in a phosphoprotein: mass mapping by combining specific enzymatic degradation with matrix-assisted laser desorption/ionization mass spectrometry. Anal Biochem 219:9–20

Meyer HE, Hoffmann-Posorske E, Korte H, Heilmeyer LM Jr (1986) Sequence analysis of phosphoserine-containing peptides. Modification for picomolar sensitivity. FEBS Lett 204:61–66

Neubauer G, Mann M (1999) Mapping of phosphorylation sites of gel-isolated proteins by nanoelectrospray tandem mass spectrometry: potentials and limitations. Anal Chem 71:235–242

Neville DC, Rozanas CR, Price EM, Gruis DB, Verkman AS, Townsend RR (1997) Evidence for phosphorylation of serine 753 in CFTR using a novel metal-ion affinity resin and matrix-assisted laser desorption mass spectrometry. Prot Sci 6:2436–2445

Resing KA, Johnson RS, Walsh KA (1995) Mass spectrometric analysis of 21 phosphorylation sites in the internal repeat of rat profilaggrin, precursor of an intermediate filament associated protein. Biochemistry 34:9477–9487

Soskic V, Gorlach M, Poznanovic S, Boehmer FD, Godovac-Zimmermann J (1999) Functional proteomics analysis of signal transduction pathways of the platelet-derived growth factor beta receptor. Biochemistry 38:1757–1764

Taylor JA, Walsh KA, Johnson RS (1996) Sherpa: a Macintosh-based expert system for the interpretation of electrospray ionization LC/MS and MS/MS data from protein digests. Rapid Commun Mass Spectrom 10:679–687

van der Geer P, Hunter T (1994) Phosphopeptide mapping and phosphoamino acid analysis by electrophoresis and chromatography on thin-layer cellulose plates. Electrophoresis 15:544–554

Watts JD, Affolter M, Krebs DL, Wange RL, Samelson LE, Aebersold R (1994) Identification by electrospray ionization mass spectrometry of the sites of tyrosine phosphorylation induced in activated Jurkat T cells on the protein tyrosine kinase ZAP-70. J Biol Chem 269:29520–29529

Wilm M, Mann M (1996) Analytical properties of the nanoelectrospray ion source. Anal Chem 68:1–8

Wilm M, Neubauer G, Mann M (1996) Parent ion scans of unseparated peptide mixtures. Anal Chem 68:527–533

Wurgler-Murphy SM, Saito H (1997) Two-component signal transducers and MAPK cascades. TIBS 22:172–176

Yates JR III, Eng JK, McCormack AL, Schieltz D (1995) Method to correlate tandem mass spectra of modified peptides to amino acid sequences in the protein database. Anal Chem 67:1426–1436

11 Glycoproteomics: High-Throughput Sequencing of Oligosaccharide Modifications to Proteins

Pauline M. Rudd, Cristina Colominas, Louise Royle, Neil Murphy, Edmund Hart, Anthony H. Merry, Holger F. Heberstreit, and Raymond A. Dwek

11.1 Beyond the Genome and the Proteome Lies the Glycome

Genomics establishes the relationship between biological processes and gene activity. Proteomics (James 1997), which relates biological activity to the proteins expressed by genes, is fundamental to our understanding of biology. It is the proteins, rather than the genes that encode them, which engage in biological events (Wilkins et al. 1995). Furthermore, most proteins contain post-translational modifications which are the products of enzyme reactions. Since the enzymes are coded for by different genes, the complete structure of an individual protein cannot be determined by reference to either a single gene or the protein sequence alone. One of the most common ways that a protein is modified is by the process of glycosylation, in which oligosaccharides are attached to specific sites encoded in the primary sequence of the protein (Dwek 1996).

11.1.1 The Importance of Glycosylation in Proteomics

It is only when a glycoprotein is viewed in its entirety that the full significance of glycosylation can be appreciated, because oligosaccharides play many varied roles in the structure and function of proteins (Rudd and Dwek 1997; Van den Steen et al. 1998). For example, proteins using the calnexin pathway to achieve correct folding must contain specific monoglucosylated oligomannose sugars. Proteins with sugars carrying the sialyl Lewis-X epitope can act as recognition markers, allowing cells to marginate at sites of inflammation. As a consequence of their large size relative to the protein domains to which they are attached, sugars shield significant areas of proteins from proteases and play a role in the spacing and orientation of cell-surface proteins.

Oligosaccharide processing is an integral part of the secretory pathway and is highly sensitive to alterations in the biological processes taking place inside the cell. For example, glycan processing can alter with disease. In rheumatoid arthritis, the levels of fully galactosylated sugars decrease with disease activity (Parekh et al. 1985). Glycan processing can also change as cells differentiate. For example, when intestinal crypt cells differentiate into mature epithelial cells, an O-glycosylated form of lactase–phlorizin hydrolase is generated (Naim and Lentze 1992). Therefore, monitoring the glycan profile of a glycoprotein can be a sensitive indicator of changing cellular events.

In proteomics, two-dimensional (2D) gel electrophoresis is used to separate and characterise both glycosylated and non-glycosylated proteins in a tissue or cell. The technique is particularly useful for comparing protein features derived from related biological samples. For example, by comparing the 2D gels of diseased tissues with those from normal controls, specific proteins unique to the diseased tissues that suggest potential therapeutic targets or diagnostics can be identified (Hochstrasser and Tissot 1993). However, when interpreting the gels, it is important to be aware that cellular changes can induce glycosylation changes, which can lead to significant shifts in the pI or molecular weights of disease-related glycoproteins. More than 60% of all natural proteins are glycosylated; therefore, a concept such as proteomics, which sets out to relate protein expression and activity to biologically significant changes in complex biological systems, must also take protein glycosylation into account.

While sequencing a linear peptide chain at the nanogram level is straightforward and rapid, the identification of the heterogeneous glycans attached to the polypeptide requires novel approaches to achieve rapid oligosaccharide analysis. The technology described here provides a rapid and robust method of N-glycan analysis that can be applied routinely to low picomoles of glycoproteins in gels, with a minimal requirement for specialised equipment or expertise. We also describe the progress we have made in the analysis of O-glycans released from femtomoles of lyophilised glycoproteins.

11.1.2 High-Throughput Analysis of Heterogeneous Glycans Requires New Analytical Strategies

Glycoproteins generally exist as populations of glycosylated variants (glycoforms) of a single homogeneous polypeptide. Although the same glycosylation machinery is available to all proteins that enter the secretory pathway in a given cell, most glycoproteins emerge with characteristic glycosylation patterns and heterogeneous populations of glycans at each glycosylation site. It is not uncommon for a glycoprotein to be processed with more than 100 alternative glycans at a single glycosylation site (Rudd et al. 1997a).

Sequencing a linear peptide chain at the nanogram level is straightforward, and amino acid sequencing is routinely available to protein chemists. In contrast, the heterogeneity and the branched structures of sugars attached to a peptide chain make the glycan analysis of a glycoprotein more complicated and inevitably require more material. The robust, rapid, automated technology for oligosaccharide sequencing that we describe here has been developed only recently and is still being refined to achieve the high sensitivity required to analyse sub-picomole levels of glycoproteins in 2D gels. The demand for rapid glycan sequencing to complement high-throughput protein sequencing requires new computer programmes, the first of which, named "PeakTime" is discussed in Section 11.2.5.

11.2 A Rapid and Robust Strategy for N- and O-Glycan Analysis

In the first instance, the primary sequence of the protein determines whether or not it will be modified by the addition of N- or O-linked oligosaccharides. N-linked sugars (Fig. 11.1A) are added to the amide side-chains of some Asn residues that form part of the triplet AsnXaaSer/Thr, while O-linked sugars (Fig. 11.1B) are most commonly found attached to the hydroxyl side-chains of some Ser or Thr residues.

The steps involved in oligosaccharide analysis are discussed below and involve releasing the sugars from the protein (Sect. 11.2.1) and labelling (Sect. 11.2.2), separating (Sect. 11.2.3) and characterising (Sect. 11.2.4) the components of the glycan pool. A straightforward strategy applicable to both N- and O-linked glycans is shown in Fig. 11.2.

11.2.1 Release of N- and O-Linked Glycans from Glycoproteins

11.2.1.1 Chemical Release of N- and O-Glycans from Glycoproteins Using Anhydrous Hydrazine

When sufficient material is available, hydrazinolysis is a general method for chemically and non-selectively releasing all N- and O-glycans from glycoproteins (Patel and Parekh 1994). The analysis of the N-glycans of immunoglobulin G (IgG) released by automated hydrazinolysis [GlycoPrep 1000; Oxford GlycoSciences (OGS), Abingdon, UK] is discussed below.

Hydrazine release of O-glycans is best achieved by manual hydrazinolysis, because the conditions can be optimised to minimise the "peeling" of oligosaccharides associated with the degradation of the terminal monosaccharides [N-acetyl galactosamine (GalNAc), in the case of O-glycans] when they are substituted at the 3 position. The release and analysis of the O-glycans from human neutrophil gelatinase B is discussed in Section 11.3.5.

11.2.1.2 Enzymatic Release and Analysis of N-Linked Oligosaccharides from Protein Bands on Sodium Dodecyl Sulphate Polyacrylamide-Gel Electrophoresis Gels

An "in-gel" release method has been developed to release N-glycans directly from protein bands on sodium dodecyl sulphate polyacrylamide-gel electrophoresis (SDS-PAGE) gels using peptide-N-glycosidase F

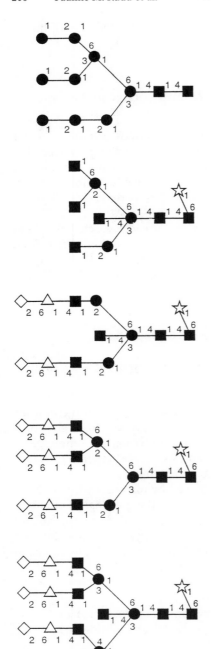

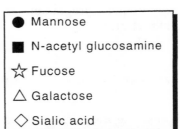

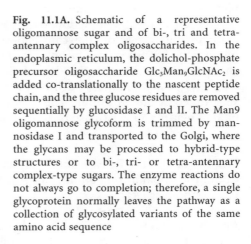

Fig. 11.1A. Schematic of a representative oligomannose sugar and of bi-, tri and tetra-antennary complex oligosaccharides. In the endoplasmic reticulum, the dolichol-phosphate precursor oligosaccharide $Glc_3Man_9GlcNAc_2$ is added co-translationally to the nascent peptide chain, and the three glucose residues are removed sequentially by glucosidase I and II. The Man9 oligomannose glycoform is trimmed by mannosidase I and transported to the Golgi, where the glycans may be processed to hybrid-type structures or to bi-, tri- or tetra-antennary complex-type sugars. The enzyme reactions do not always go to completion; therefore, a single glycoprotein normally leaves the pathway as a collection of glycosylated variants of the same amino acid sequence

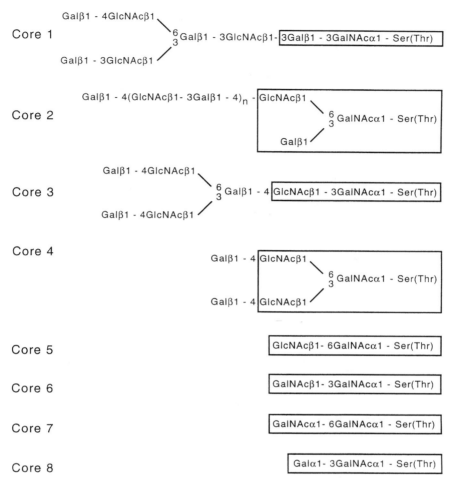

Fig. 11.1B. Schematic representation of the structures of the common *O*-glycan cores. *O*-glycans are attached by the stepwise addition of single monosaccharides, normally beginning with *N*-acetylgalactosamine. The common core structures may be extended to form the backbone region by the addition of galactose (linked β1–3/4) and GlcNAc (linked β1–3/6). Terminal residues include galactose, *N*-acetylgalactosamine, fucose, sialic acid and sulphate

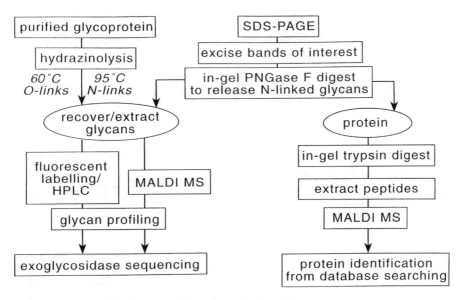

Fig. 11.2. Strategies for glycan analysis and protein identification

(PNGase F) (Küster et al. 1997). The same enzyme, which hydrolyses the β-aspartylglycosylamine bond between asparagine and the innermost GlcNAc of the glycan, can also be used to release N-glycans from glycoproteins in solution. At present, there is no corresponding strategy to release O-glycans, because a generic O-glycanase that can release all O-linked sugars from a protein remains to be identified. Since hydrazinolysis cannot be used on gels, further technical innovations are needed to allow the analysis of O-glycans attached to proteins in 2D gels.

Once released, the N-glycan pool can be analysed in the same way a pool released by hydrazinolysis is analysed: directly by matrix-assisted laser desorption ionisation time-of-flight mass spectrometry (MALDI-TOF MS) and (after fluorescent labelling) by high-performance liquid chromatography (HPLC). The preliminary assignments made to components of the glycan pool should be confirmed by analysing the products of digestions of the pool with enzyme arrays (Guile et al. 1996), either by MS or HPLC.

11.2.1.3 Advantages of the "In-Gel" Release Method

The development of the "in-gel" method has opened the way to analysing glycans directly from proteins separated on the 2D gel systems used in proteomics. An advantage of the "in-gel" release method is that the protein remains in the gel after the sugars have been removed. Subsequently, "in-gel" proteolysis of the protein with trypsin and analysis of the peptide fragments by nanospray MS enables the protein to be identified. This is achieved by using a protein database to compare the molecular weights of the tryptic fragments with those of the predicted sequences of fragments from tryptic digests of known proteins (Küster et al. 1997). The ability to release sugars directly from SDS-PAGE gels eliminates the need for extensive protein purification of molecules that are difficult to isolate or that are available only in limited amounts. For example, when gelatinase B from neutrophils is purified on gelatin–Sepharose, it co-purifies with its natural inhibitor, tissue-inhibitor metalloproteinase-1 (TIMP-1; Opdenakker et al. 1991). Although this enzyme-inhibitor complex is stable, SDS-PAGE can be used to resolve the two proteins. N-linked oligosaccharides can then be released directly from the gel band containing gelantinase B (Rudd et al. 1997b).

Another application of the technology is in situations where reducing gels allow the straightforward separation of proteins into their component sub-units. Examples include the resolution of IgG into heavy and light chains and the surface-coat proteins of the hepatitis B sub-viral particle. These can be separated into three major glycoprotein components, small, middle and large (S, M and L, respectively) by SDS-PAGE analysis under reducing conditions (Mehta et al. 1997). The "in-gel" release method can also be used to release the sugars from peptide fragments resulting from protease digests when these can be resolved by SDS-PAGE, leading to rapid glycosylation-site analysis.

11.2.2 Labelling the Glycan Pool

N- and O-linked oligosaccharides released by hydrazine, or N-linked glycans released by PNGase F, exist in an equilibrium between the cyclic- (hydroxyl at C1) and opened-ring forms (aldehyde at C1) of the reducing terminal GlcNAc or GalNAc residue. The opened-ring form can be derivatised at C1 by fluorophores, such as 2-aminobenzamide (2-AB), via a reductive amination reaction that initially forms a Schiff's base with the sugar. The reaction is driven to completion by the reduction of the Schiff's base to an amine functional group, thereby favouring the forward direction of the tautomerism (Fig. 11.3).

Although MS can be applied to unlabelled sugars, in most other techniques, the ability to label the reducing termini of the sugars can be exploited to increase the sensitivity of detection. A range of fluorescent molecules that allow detection of sugars in the femtomole range is available. A suitable label must have a high molar-labelling efficiency and must label the oligosaccharide components of a glycan pool non-selectively so that they are detected in their correct molar proportions. 2-AB fulfils these requirements and is compatible with a range of separation techniques, including normal-phase (NP), reverse-phase (RP) and anion-exchange HPLC, MALDI-TOF, electrospray MS and BioGel P4 gel-permeation chromatography (Bigge et al. 1995). HPLC analysis of 2-AB labelled glycans requires only 5% of the material needed for one MALDI-MS run. Other labelling systems that will retain all these advantages

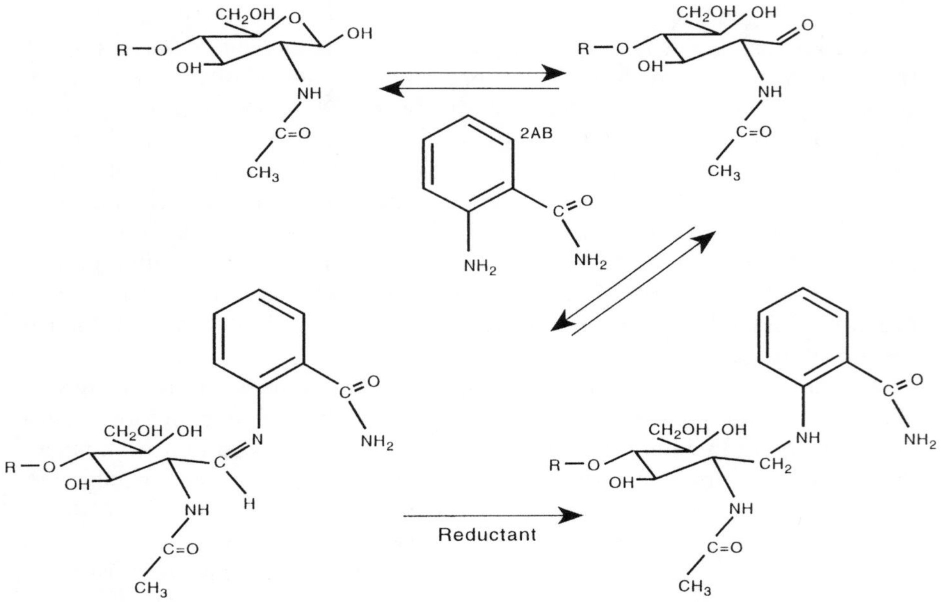

Fig. 11.3. Mechanism for the reductive amination of an oligosaccharide with the fluorophore 2-aminobenzamide. The figure indicates that the reaction can only take place when the terminal GlcNAc residue in the chitobiose core is in the "ring-opened form". R indicates the remainder of the oligosaccharide

but give the significantly increased sensitivity needed to analyse the glycosylation of minor components of 2D gels must be developed.

11.2.3 Resolving Released Glycan Pools and Assigning Structure by HPLC

Three types of HPLC matrices are commonly used for separating released oligosaccharides into pools according to their different characteristics. A major advantage of all HPLC systems is that they can be used preparatively. This means, for example, that the glycans in individual peaks can be collected, the volatile buffers removed, and the glycans analysed by MS or capillary electrophoresis to validate assignments.

11.2.3.1 Weak Anion-Exchange Chromatography

Weak anion-exchange chromatography separates sugars on the basis of the number of charged groups they contain. For example, Fig. 11.4 shows the separation of a library of (1) N- and O-linked sugars released from CD59 and (2) N-linked sugars from fetuin. In the first instance, this method separates

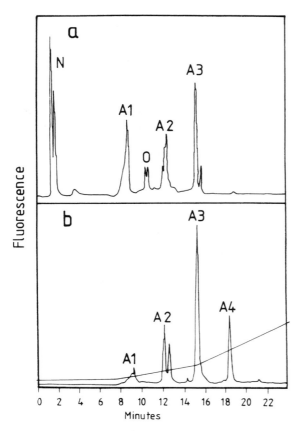

Fig. 11.4a,b. Weak anion-exchange analysis of (**a**) CD59 and (**b**) fetuin glycans on the basis of charge. *A1*, *A2* and *A3* represent mono-, di- and tri-sialylated *N*-links, respectively. *N* neutral glycans, *O* mono-sialylated *O*-linked glycans. The analysis was performed on a Vydac 301VHP575 polymer-matrix column (Guile et al. 1994). The conductivity of the gradient (solvent A: ammonium formate, pH 9; solvent B: water) is shown as a line in panel (**b**)

the mixture according to charge; however, within one charge band, larger structures elute before smaller ones. The double peaks within the mono- and di-sialylated fetuin bands contain tri- and bi-antennary structures in which the tri-antennary glycans elute first.

11.2.3.2 Normal-Phase HPLC

NP-HPLC is performed using a column with amide functional groups that interact with the hydroxyl groups on the sugars. On such a column, glycans are resolved on the basis of hydrophilicity. Sugars applied in high concentrations of an organic mobile-phase adsorb to the column surface and are subsequently eluted by an aqueous gradient. In general, larger oligosaccharides are more hydrophilic than small ones and require higher concentrations of aqueous solvent to elute them.

A sensitive and reproducible NP-HPLC technology has been developed (Guile et al. 1996) using the GlycoSep-N column (Glyko Inc.). This system is capable of resolving sub-picomole quantities of mixtures of fluorescently labelled neutral and acidic N- and O-glycans simultaneously and in their correct molar proportions. Elution positions are expressed as glucose units (gu) by comparison with the elution positions of glucose oligomers (dextran ladders). For N-linked sugars, the contribution of individual monosaccharides to the overall gu value of a given glycan can be calculated, and these incremental values are used to predict the structure of an unknown sugar from its gu value. The GlycoSep-N column is able to resolve arm-specific substitutions of galactose and can resolve the linkage position, which can also affect retention time; thus, it provides a further level of specificity (Guile et al. 1996). The resolving power and reproducibility of predictive NP-HPLC, and the ability to analyse sialylated and neutral sugars in one run, make it particularly useful for profiling both N- and O-glycan pools and for making comparisons between different samples. An example of this is the comparison of the glycosylation of a range of leukocyte antigens all expressed in the same Chinese hamster ovary (CHO) cell line (Fig. 11.5; Rudd et al. 1999a).

11.2.3.3 Reverse-Phase HPLC

RP-HPLC using C-18 columns (GlycoSep R: Glyko Inc.) separates sugars on the basis of hydrophobicity. Samples applied in aqueous buffer are eluted with increasing concentrations of organic solvent. Sugars that co-elute on NP-HPLC can often be separated by RP-HPLC (Fig. 11.6). This can be particularly useful for distinguishing between bisected and non-bisected structures, which elute very close to each other on NP-HPLC but are well resolved on RP. The standard dextran ladder is not well resolved with the RP gradient; therefore, the elution positions of sugars separated by RP are measured in comparison with an arabinose ladder and are assigned arabinose units (au).

11.2.4 Simultaneous Sequencing of Oligosaccharides Using Enzyme Arrays

The preliminary assignment of structures from the initial NP-HPLC run can be confirmed rapidly by sequencing all of the oligosaccharides in a glycan pool simultaneously using enzyme arrays. Enzymatic analysis of oligosaccharides using highly specific exoglycosidases is a powerful means of determining the sequences and structures of glycan chains. However, until recently, it was necessary to isolate single sugars from the glycan pool for digestion with exoglycosidases, either sequentially or in arrays. The high resolving power of the NP-HPLC system allowed a new approach to be developed (Guile et al. 1996).

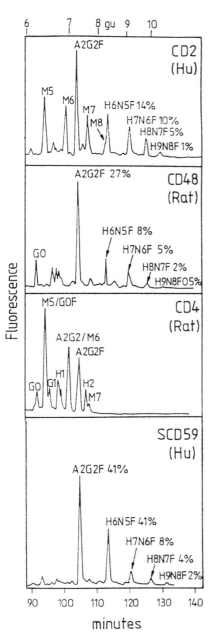

Fig. **11.5.** Normal-phase high-performance liquid-chromatography analysis of recombinant, soluble forms of cell-surface antigens expressed in Chinese-hamster ovary cells The analysis was performed on GlycoSep N column using a gradient-elution system (solvent A: 50 mm ammonium formate, pH 4.4; solvent B: acetonitrile)

Normal Phase Gu values   Reverse Phase Au values

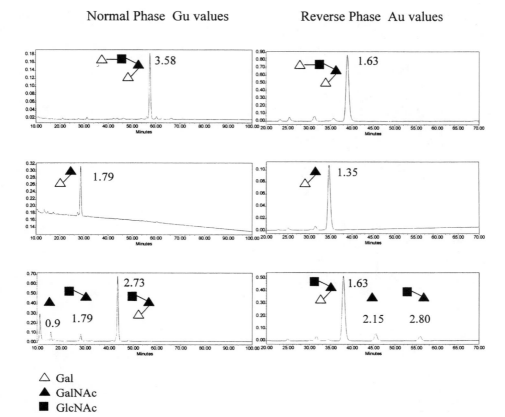

△ Gal
▲ GalNAc
■ GlcNAc
＼ β1–6
— β1–4
╱ β1–3

Fig. 11.6. Normal-phase (NP) and reverse-phase (RP) high-performance liquid-chromatography analysis of O-linked sugars. NP analysis was performed on a GlycoSepN column [Glyko Inc.] using a gradient of 20–58% solvent A during a 152-min period (solvent A: 50-mM ammonium formate, pH 4.4; solvent B: acetonitrile). The RP analysis was performed on a GlycoSep R column (Glyko Inc.) using a gradient of 95–76% solvent A during a 165-min period (solvent A: 50-mM triethylamine/formic acid, pH 5; solvent B: 50% solvent A, 50% acetonitrile). The NP glucose-unit values are derived from a dextran ladder, and the RP arabinose-unit values are derived from an arabinose ladder. The glycans Galβ1–3GalNAc and GlcNAcβ1–6GalNAc co-elute with NP but are separated by RP, whereas Galβ1–4GlcNAcβ1–6(Galβ1–3)GalNAc and GlcNAcβ1–6(Galβ1–3)GalNAc co-elute with RP but are separated by NP

This involves the simultaneous analysis of the total glycan pool by digesting aliquots with a set of enzyme arrays. After overnight incubation, the products of each digestion are analysed by NP-HPLC or MALDI-TOF MS. On the HPLC system, structures are assigned to each peak from the known specificities of the enzymes and the pre-determined incremental values of individual mono-saccharide residues. To illustrate this technique, the rapid profiling and simul-taneous analysis of the major *N*-glycans attached to rat CD48 expressed in CHO cells is shown in Fig. 11.7.

This strategy is currently being extended to the analysis of *O*-linked glycans, such as those attached to human neutrophil gelatinase B. The NP-HPLC profile of the total glycan pool is shown in Fig. 11.8 (Rudd et al. 1999b). The subsequent analysis using enzyme arrays indicated that human neutrophil gelatinase B contains mainly *O*-linked glycans with core type 1 (Galβ1–3GalNAcα1-R).

11.2.5 Analysis of the NP-HPLC Results

The increasing use of HPLC columns for oligosaccharide analysis has coin-cided with a shift in emphasis from biochemical analysis at the bench towards post-HPLC graphical or numerical analysis. Behind this shift is the recent move towards the simultaneous analysis of whole glycan pools; this has many advan-tages but also increases the complexity of the later analysis. In effect, the separation of individual glycans, previously performed at the bench, is now performed conceptually.

There are several steps in the analysis. NP-HPLC resolves the released and 2-AB labelled glycan pool. Each peak in the output log is then calibrated with reference to a standard dextran ladder (gu values; x axis, Fig. 11.8) using a poly-nomial line fit. Initial assignments are made by using tables containing exper-imentally determined gu values of known standards and generic incremental values, which describe the addition of single monosaccharide residues to standard glycan cores. These values are highly reproducible (±0.02 gu).

The entire glycan pool is then digested with different exoglycosidase arrays, and the products are resolved on the same NP-HPLC system. The calibrated chromatograms of the original pool and the digests are then displayed verti-cally, one below the other, all to the same horizontal scale (measured in gu). The consistent use of this single scale is important, because it allows direct comparison of the different runs and, in particular, it means that shifts of indi-vidual peaks due to the action of a single enzyme appear to be of equal length everywhere on the axis. The identification of individual peaks is deduced by inspection of the successive shifts, because each exoglycosidase enzyme removes a further part of the glycan chain. This process confirms the initial assignment of individual peaks from their gu or au values.

Previously, the analysis was usually performed graphically, by inspection; however, a software package, PeakTime, which will automate much of the process, is presently under construction. At present, PeakTime is capable of performing the initial calibration step and is used daily for this purpose in the lab. Without the aid of a computer, it would be virtually impossible to redraw every chromatogram to the calibrated scale that facilitates their comparison. In the near future, a database containing the elution positions of known

Sequencing CD48 glycans

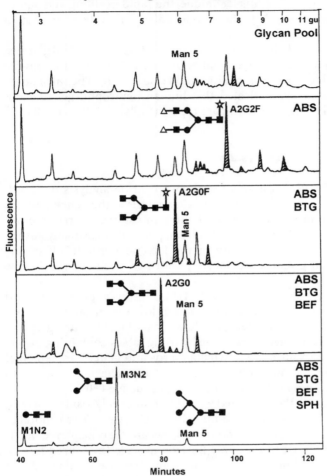

Fig. 11.7. Simultaneous analysis of *N*-glycans released from rat CD48 (expressed in Chinese-hamster ovary cells) using enzyme arrays. The figure shows the high-performance liquid-chromatography analysis of the total glycan pool (*top panel*) and the products resulting from the digestion of four aliquots of the total CD48 glycan pool with a series of enzyme arrays. The particular enzyme array that produced each profile is shown in the appropriate panel. The *shaded areas* define peaks that contain glycans that were subsequently digested by the additional enzyme present in the next array. The glucose-unit (gu) value of each peak was calculated by comparison with the dextran-hydrolysate ladder shown at the *top of the figure*. Structures were assigned from the gu values, previously determined incremental values for monosaccharide residues (Guile et al. 1996) and the known specificity of the exoglycosidase enzymes. The structures of the most abundant glycan populations are shown. *ABS Arthrobacter ureafaciens* sialidase, *BEF* bovine-epididymis α-fucosidase, *BTG* bovine-testes β-galactosidase, *SPH Streptococcus pneumoniae* β-*N*-acetylhexosaminidase

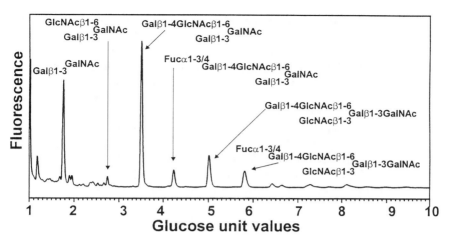

Fig. 11.8. NP-HPLC profile of de-sialylated *O*-linked oligosaccharides from gelatinase B. The structures were assigned from their glucose-unit values and confirmed by exoglycosidase digestions. (Rudd et al. 1999b)

glycans and monosaccharide incremental values will be centrally available, kept up to date and accessed rapidly and conveniently from the same software suite that is used in the analysis itself. The specificities of enzymes will also be stored in the database, enabling the rapid identification of the associated shifts.

11.3 Applications of this Technology

In the examples below, we demonstrate how rapid oligosaccharide sequencing of fluorescently labelled sugars and the use of an oligosaccharide database that gives the dimension of the sugars can be combined with protein-structure data to give insights into roles for glycosylation in the function of individual proteins.

11.3.1 IgG Glycosylation and Disease

Protein glycosylation is influenced by the local protein structure and by the available repertoire of glycosylating enzymes in the cell where the protein is being glycosylated. Therefore, glycosylation patterns are sensitive markers of changes within the environment of the cell. For example, galactosylation levels of IgG vary during pregnancy and in a limited number of diseases, including tuberculosis, Crohn's disease and rheumatoid arthritis (RA). In RA, the galactosylation status of IgG is a prognostic indicator, and high levels of sugars lacking galactose (G0 types) compared with levels in age-matched normal controls correlate with active and severe disease (Fig. 11.9).

The absence of the terminal galactose residue in G0-type sugars results in decreased protein–oligosaccharide interactions in the crystalline fragment and allows displacement of the sugars out of the space between the CH_2 domains. As a consequence, the exposed terminal GlcNAc residues are in a position where they can be recognised by endogenous lectins, such as mannose-binding lectin (MBL). The clustering of the G0-type glycoforms may contribute to the pathogenesis of RA by providing a mechanism for the activation of the classical complement pathway via multivalent binding to MBL (Malhotra et al. 1995). This is expected to be important in the synovial cavity, where MBL levels are increased (Malhotra et al. 1995) and G0 levels are elevated (Parekh et al. 1985).

11.3.2 Potential Roles for the Glycans Attached to CD59

Host cells are normally protected from lysis by CD59, a cell-surface glycoprotein that binds to the complement proteins C8 and/or C9 in the nascent membrane attack complex. CD59 belongs to the Ly-6 superfamily and is present on a wide variety of cell types, including leukocytes, platelets, epithelial and endothelial cells, placental cells and erythrocytes. CD59 is attached to the surfaces of these cells by means of a glycosylphosphatidyl inositol (GPI) anchor containing three lipid chains (Rudd et al. 1997b). Human erythrocyte CD59 consists of a heterogeneous mixture of more than 120 glycoforms (Fig. 11.10A), of which the major single sugar is a complex glycan containing both a bisecting GlcNAc residue and a core fucose (Rudd et al. 1997b).

Figure 11.10B shows the potential for combining proteomics with glycan analysis. A molecular model of CD59 was constructed using data from glycan analysis of the sugars and the glycosylphosphatidyl anchor (Rudd et al. 1997a), the oligosaccharide structural database, and the protein coordinates from the solution structure (Fletcher et al. 1994). The N-linked oligosaccharides (size range 3–6 nm), which are attached to the disc-like extracellular region of CD59 (diameter approximately 3 nm), project from the protein domain in the plane of the active face and adjacent to the membrane surface. From a structural point of view, CD59 is the most well-defined cell-surface molecule known.

The ability to view the molecule in its entirety makes it possible to suggest some possible roles for the sugars. The heterogeneous glycans are expected to influence the geometry of the packing, and it is likely that they will also prevent the aggregation of CD59 molecules on the cell surface. By limiting non-specific protein–protein interactions and controlling the protein spacing, the glycans may influence the distribution of CD59 molecules at the cell surface, where GPI-anchored proteins may associate in dynamic equilibrium with isolated individual molecules in micro-domains (Van den Berg et al. 1995). The large N-glycans may also be important in preventing proteolysis of the extracellular

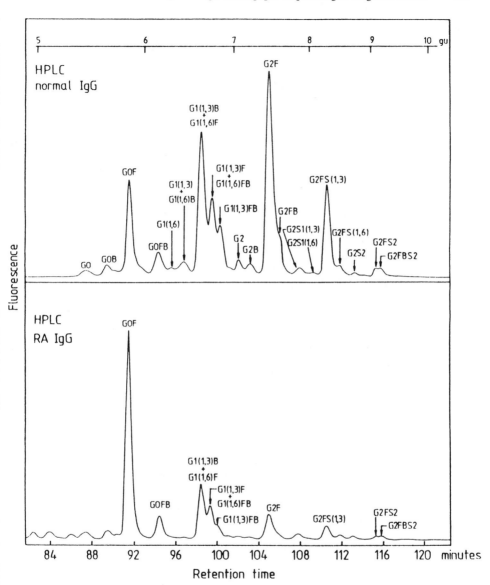

Fig. 11.9. NP-HPLC separation of total immunoglobulin G (IgG) glycans. The glycans were released from (*above*) pooled normal-serum IgG and (*below*) rheumatoid-serum IgG from a single patient by hydrazinolysis. The sugars were fluorescently labelled and resolved by NP-HPLC. Peaks were assigned glucose-unit (gu) values by comparison with the elution positions of a standard dextran ladder (shown at the *top of the figure*). The peaks were assigned structures from their gu values, exoglycosidase digestions and from the results of co-injection of standard glycans (Guile et al. 1996). The percentage of bi-antennary sugars in which both arms lack galactose and terminate in GlcNAc (G0 type) was 51.8% in rheumatoid arthritis IgG, compared with 15.2% in normal-serum IgG

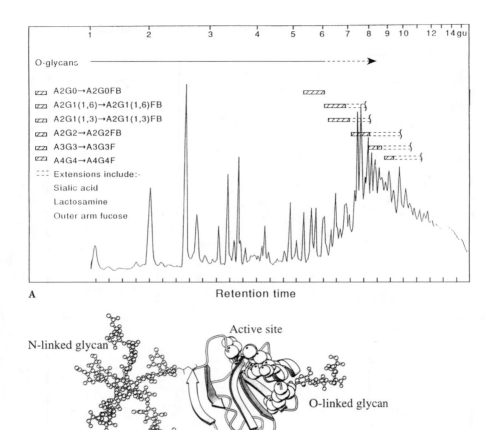

A

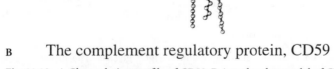

B   The complement regulatory protein, CD59

Fig. 11.10. **A** Glycosylation profile of CD59. **B** A molecular model of CD59, based on the protein co-ordinates from the solution structure. The binding-site residues (*large white spheres*) are located at Trp40, Asp24, Arg53 and Glu56 (Bodian et al. 1997). The glycan anchor is modelled with a tri-mannosyl core, an ethanolamine bridge at Man3 and additional ethanolamine groups at Man1 and Man2. Two lipids are attached to inositol via phosphate, and the third is attached directly to the inositol ring by an ester linkage. A trisialylated, tetra-antennary complex *N*-glycan is shown attached to Asn 18. The *O*-glycan, NeuNAcα2,3Galβ1–3GalNAc, is attached to Thr51 to indicate one of the possible linkage positions

domain, because *N*-glycosylation has been shown to increase the dynamic stability of a protein, while different glycoforms variably increase its resistance to protease digestion (Rudd et al. 1999a).

11.3.3 The Three-Dimensional Structure of the Protein Influences Glycan Processing

Most cell-surface molecules are glycoproteins consisting of linear arrays of globular domains that contain stretches of amino acid sequences similar to regions in other proteins. These conserved regions form the basis for the classification of proteins into superfamilies. Recombinant, soluble forms of six leukocyte antigens belonging to the Ly-6 (human sCD59), scavenger-receptor (human sCD5) and Ig (human sCD2, rat sCD48, rat sCD4 and rat sThy1) superfamilies were expressed in the same CHO cell line. Figure 11.5 compares the glycan profiles of some these proteins, showing that *N*-linked oligosaccharide processing is different for each and indicating that it depends on the overall tertiary structures of the molecules. Thus, the surface of a single cell will display a diverse repertoire of glycans on its range of glycoproteins. This precludes the presentation of multiple copies of a single oligosaccharide on the cell surface, which might make the cell vulnerable (for example) to bacteria, which bind to specific sugars that are multiply presented.

11.3.4 Simultaneous Analysis of the *N*-Glycan Pool from Rat CD48 Expressed in CHO Cells Reveals Extensive Heterogeneity

Soluble rat CD48, containing five *N*-linked glycosylation sites, was expressed in CHO cells. The oligosaccharides in the glycan pool were analysed simultaneously using enzyme arrays (Fig. 11.7; Rudd et al. 1999a). In molecular-modelling studies, the range of complex, hybrid and oligomannose structures were found to shield large areas of the two IgSF domains of which CD48 is composed (Rudd et al. 1999a). CD48 is the major ligand for CD2, and this cell-adhesion pair mediates the precise alignment of the cell surfaces of cytolytic T-lymphocytes carrying the T-cell receptor complex with those of target cells carrying loaded human-leukocyte antigen class-1 molecules (Davis and Van der Merwe 1996; Dustin et al. 1996). The glycosylation analysis suggests that the sugars protect CD48 against proteases and play a role in the packing of CD48 on the cell surface. In addition, the sugars at the sites close to the membrane may serve to orient the face of the protein that contains the binding sites.

11.3.5 Analysis of the *O*-Glycans from Human Neutrophil Gelatinase B Suggests That They May Produce an Extended and Rigid Region of the Peptide

Gelatinase B is a multi-domain metalloproteinase that cleaves extracellular-matrix substrates, such as denatured collagens (gelatins), after these have been clipped by collagenases, stromelysin or other metalloproteinases. Gelatinase B contains seven protein domains, three potential *N*-glycosylation sites, a Pro/Ser/Thr-rich region and a number of isolated Ser and Thr residues. The *O*-glycans, which constitute more than 85% of the total sugars, are expected to be located mainly in the Ser/Thr/Pro-rich domain. The *O*-glycans were released by optimised manual hydrazinolysis, profiled (Fig. 11.8) and analysed using enzyme arrays (Rudd et al. 1999b). The glycans in the major peaks contained core types 1, 2 and 6. The attachment of these glycans to the Ser/Thr/Pro-rich domain is expected to produce an extended and rigid region of the peptide. Electron-microscopic studies indicate that the extension contributed per residue in an *O*-glycosylated peptide is approximately 0.25 nm in mucins (Shogren et al. 1989).

11.4 Conclusion

The technology described here has been developed to enable *N*- and *O*-linked sugars released from proteins to be analysed rapidly and routinely. The minimum need is for an HPLC system equipped with a NP column and a fluorescence detector, a series of glycan standards, and a set of endo- and exo-glycosidase enzymes. MS provides valuable complementary information. While MS cannot give information about the positions or linkages of mono-saccharides within an oligosaccharide, it enables peak assignments to be based on a combination of their molecular weighte and their elution positions on HPLC. The ability to sequence *N*-linked sugars directly from gels has eliminated the need for exhaustive protein purification, which will in the future make feasible the glycosylation analysis of biological samples available only in low yields from 2D gels. The aim of developing such a strategy is to enable glycoproteins to be viewed in their entirety so that, within the proteome project, more insight can be gained into the complementary roles sugars and proteins play in the structure and function of glycoproteins.

References

Bigge JC, Patel TP, Bruce JA, Goulding PN, Charles SM, Parekh RB (1995) Nonselective and efficient fluorescent labeling of glycans using 2-amino benzamide and anthranilic acid. Anal Biochem 230:229–238

Bodian DL, Davis SJ, Rushmere NK, Morgan BP (1997) Mutational analysis of the active site and antibody epitopes of the complement-inhibitory glycoprotein, CD59. J Exp Med 185:507–516

Davis SJ, van der Merwe PA (1996) The structure and ligand interactions of CD2: implications for T-cell function. Immunol Today 17:177–187

Dustin ML, Ferguson LM, Chan PY, Springer TA, Golan DE (1996) Visualization of CD2 interaction with LFA-3 and determination of the two-dimensional dissociation constant for adhesion receptors in a contact area. J Cell Biol 132:456–474

Dwek RA (1996) Glycobiology: toward understanding the function of sugars. Chem Rev 96:683–720

Fletcher CM, Harrison RA, Lachman PJ, Neuhaus D (1994) Structure of a soluble, glycosylated form of the human complement regulatory protein CD59. Curr Biol Struct 2:185–199

Guile GR, Wong SYC, Dwek RA (1994) Analytical and preparative separation of anionic oligosaccharides by weak anion-exchange high-performance liquid chromatography on an inert polymer column. Anal Biochem 222:231–235

Guile GG, Rudd PM, Wing DR, Prime SB, Dwek RA (1996) A rapid high resolution method for separating oligosaccharide mixtures and analysing sugarprints. Anal Biochem 240:210–226

Hochstrasser DF, Tissot J-D (1993) Cinical applicaation of high resolution two dimensional polyacrylamide gel electrophoresis. In: Chrombach A, Dunn MJ (eds) Advances in electrophoresis, vol 6. Rudola VCH, Weinheim, pp 270–375

James P (1997) Protein identification in the post-genome era: the rapid rise of proteomics. Q Rev Biophys 30:279–331

Küster B, Wheeler SF, Hunter AP, Dwek RA, Harvey DJ (1997) Sequencing of N-linked oligosaccharides directly from protein-gels: in-gel deglycosylation followed by matrix-assisted laser desorption/ionisation mass spectrometry and normal-phase high-performance liquid chromatography. Anal Biochem 250:82–101

Malhotra R, Wormald MR, Rudd PM, Fischer PB, Dwek RA, Sim RB (1995) Glycosylation changes of IgG associated with rheumatoid arthritis can activate complement via the mannose binding protein. Nat Med 1:237–241

Mehta A, Lu X, Block TM, Blumberg BS, Dwek RA (1997) Hepatitis B virus envelope proteins vary drastically in their sensitivity to glycan processing. Proc Natl Acad Sci USA 94:1822–1827

Naim HY, Lentze MJ (1992) Impact of O-glycosylation on the function of human intestinal lactasephlorizin hydrolase. Characterisation of glycoforms varying in enzyme activity and localisation of O-glycoside addition. J Biol Chem 267:25494–25504

Opdenakker G, Masure S, Proost P, Billiau A, Van Damme J (1991) Natural human monocyte gelatinase and its inhibitor. FEBS Lett 284:73–78

Parekh RB, Dwek RA, Sutton BJ, Fernandes DL, Leung A, Stanworth D, Rademacher TW, Mizuochi T, Taniguchi T, Matsuta K, Takeuchi F, Nagano Y, Miyamoto T, Kobata A (1985) Association of rheumatoid arthritis and primary osteoarthritis with changes in the glycosylation pattern of total serum IgG. Nature 316:452–457

Patel TP, Parekh RB (1994) Release of oligosaccharides from proteins by hydrazinolysis. Methods Enzymol 230:57–66

Rudd PM, Dwek RA (1997) Glycosylation: heterogeneity and the 3D structure of the protein. Crit Rev Biochem Mol Biol 32:1–100

Rudd PM, Morgan BP, Wormald MR, Harvey DJ, van den Berg CW, Davis SJ, Ferguson MAJ, Dwek RA (1997a) The glycosylation of the complement regulatory protein, CD59, derived from human erythrocytes and human platelets. J Biol Chem 272:7229–7244

Rudd PM, Guile GR, Küster B, Harvey DJ, Opdenakker G, Dwek RA (1997b) Oligosaccharide sequencing technology. Nature 388:205–208

Rudd PM, Wormald MR, Harvey DJ, Devashayem M, McAlister MSB, Barclay AN, Brown MH, Davis SJ, Dwek RA (1999a) Oligosaccharide processing in the Ly-6, scavenger receptor and immunoglobulin superfamilies – implications for roles for glycosylation on cell surface molecules. Glycobiology 9:443–458

Rudd PM, Mattu TS, Masure S, Bratt T, Van den Steen PE, Wormald MR, Kuster B, Harvey DJ, Borregaard N, Van Damme J, Dwek RA, Opdenakker G (1999b) Glycosylation of natural human neutrophil gelatinase B and neutrophilgelatinase B-associated lipocalin. Biochemistry 38:13937–13950

Shogren R, Gerken TA, Jentoft N (1989) Role of glycosylation on the conformation and chain dimensions of O-linked glycoproteins: light-scattering studies of ovine submaxillary mucin. Biochemistry 28:5525–5536

Van den Berg CW, Cinek T, Hallett MB, Horejsi V, Morgan BP (1995) Exogenous glycosyl phosphatidylinositol-anchored CD59 associates with kinases in membrane clusters on U937 cells and becomes Ca(2+)-signaling competent. J Cell Biol 131:669–677

Van den Steen P, Rudd PM, Dwek, RA, Opdenakker G (1998) Concepts and principles of O-linked glycosylation. Crit Rev Biochem Mol Biol 33:151–208

Wilkins MR, Sanchez J-C, Gooley AA, Appel RD, Humphrey-Smith I, Hochstrasser, DF, Williams KL (1995) Progress with Proteome projects: why all proteins expressed by a genome should be identified and how to do it. Biotechnol Genet Eng Rev 13:19–50

12 Proteomics Databases

Hanno Langen and Peter Berndt

12.1 Introduction

Proteomics relies strongly on existing databases, such as DNA- and protein-sequence databases. At the same time, the technologies used for proteomics, such as two-dimensional (2D) electrophoresis and mass spectrometry (MS) are now generating immense amounts of data. All of this data has to be properly integrated and stored to allow easy access and searching. The data that are produced in a typical proteomics experiment can be subdivided into the following categories. The starting point is a biological sample that can be described by the origin, organ or tissue and kind of treatment or disease. The biological sample is usually analysed by one or more separation techniques. The most frequently used and effective separation techniques are 1D or 2D electrophoresis. The gels are stained then scanned, and the digital images are stored and entered into databases. Many of the existing proteomics databases store only these images and a description of the biological origin of the samples applied to the gels. For each gel, several hundred spots are recorded, and the staining intensity of each is measured. This data can be stored in another level of a database. Some commercial software tools are available to perform spot detection, gel matching and data organisation. 2D-gel technology has existed for more than 20 years (Klose 1975; O'Farrell 1975), and databases were created to track differences in protein expression.

The main challenge was the identification of the proteins representing a changing spot. During the last 5 years, new analytical tools have emerged for protein identification, and MS (Patterson and Aebersold 1995) has become the de facto tool used to identify proteins separated by gel electrophoresis. This became possible due to improvements in MS instruments, achieving higher sensitivity and accuracy. The other important developments are the fast-growing nucleotide and protein sequence databases. The identification of proteins using MS requires access to sequence databanks. The most widespread method for protein identification is the peptide-mass fingerprint (Henzel et al. 1993; James et al. 1993). The masses of peptides derived from in-gel proteolytic digestion are measured and compared with a computer-generated list formed from the simulated digestion of a protein database or a translated-nucleotide database using the same specific enzyme. Using the mass accuracy now available (~10 ppm), this technique is able to identify proteins with high confidence

(Berndt et al. 1999). For completely sequenced genomes like *Haemophilus influenzae* (Fleischmann et al. 1995), yeast (Mewes et al. 1997) or *Caenorhabditis elegans* (Chalfie 1998), the masses of all tryptic peptides from the predicted open reading frames can be calculated. The human genome (and most of the important model organisms, such as rat, mouse, zebra fish, *Drosophila*, etc.) should be completed within a few years, though many other organisms will probably never be sequenced. In January 2000, Celera Genomics completed the genome sequence of *Drosophila* and announced that the human genome was 90% complete. In the meantime, expressed-sequence tag (EST) databases and tandem MS (MS/MS)-based approaches (Yates 1998) can be used to identify proteins. Table 12.1 shows a list of completely sequenced genomes.

12.2 Protein- and Nucleotide-Sequence Databases

In this section, we describe existing nucleotide and protein databases and focus on their strengths and weaknesses. A list of the main protein and nucleotide databases is given in Table 12.2. Most of the protein and nucleotide databases are public and available via the Worldwide Web (WWW). The availability of well-annotated databases that contain certified sequence information is immensely valuable.

12.2.1 Swiss-Prot

One of the best-annotated and curated sequence databases is the Swiss-Prot database (Bairoch and Apweiler 1999). Swiss-Prot is an annotated protein-sequence database maintained collaboratively by the Department of Medical Biochemistry of Geneva and the European Bioinformatic Institute. In Swiss-Prot, as in most other sequence databases, two classes of data can be distinguished: the core data and the annotation. For each sequence entry, the core data consists of the sequence data. This data is sufficient for the identification of a protein by peptide-mass fingerprinting, but the most valuable information is the annotations and the links to other databases and to literature. The annotations describe the following topics:

- Function(s) of the protein
- Post-translational modification(s) (PTMs; carbohydrates, phosphorylation, acetylation, glycosylphosphatidyl-inositol anchors, etc.)
- Domains and sites (calcium-binding regions, adenosine triphosphate-binding sites, zinc fingers, homeoboxes, kringle, etc.)
- Secondary structure
- Quaternary structure (homodimer, heterotrimer, etc.)
- Similarities to other proteins

Table 12.1. Completed genome sequences and their sources

Organism	Size (kb)	ORF number	URL	Institution	Reference
Haemophilus influenzae KW20	1,830	1,850	http://www.tigr.org	TIGR	Fleischmann et al. (1995)
Mycoplasma genitalium G-37	580	468	http://www.tigr.org	TIGR	Fraser et al. (1995)
Synechocystis sp. PCC 6803	3,573	3,168	http://www.kazusa.or.jp	Kazusa DNA Research Institute	Kaneko et al. (1996)
Mycoplasma pneumoniae M129	816	677	http://www.zmbh.uni-heidelberg.de	University of Heidelberg	Himmelreich et al. (1996)
Escherichia coli K12-MG1655	4,639	4,289	http://www.bact.wisc.edu	University of Wisconsin ECDC	Blattner et al. (1997)
Helicobacter pylori 26695	1,667	1,590	http://www.tigr.org	TIGR	Tomb et al. (1997)
Bacillus subtilis 168	4,214	4,099	http://bioweb.pasteur.fr/GenoList/SubtiList	European and Japanese Consortium	Kunst et al. (1997)
Borrelia burgdorferi B31	1,230	1,256	http://www.tigr.org	TIGR; Brookhaven National Laboratory	Fraser et al. (1997)
Aquifex aeolicus VF5	1,551	1,544	http://www.biocat.com	Diversa Corporation; UIUC	Deckert et al. (1998)
Mycobacterium tuberculosis H37Rv (lab strain)	4,447	4,402	http://www.sanger.ac.uk	Sanger Center	Cole et al. (1998)
Treponema pallidum Nichols	1,138	1,041	http://www.tigr.org	TIGR; University of Texas	Fraser et al. (1998)
Chlamydia trachomatis serovar D	1,042	896	http://www.berkeley.edu	UC Berkeley; Stanford University	Stephens et al. (1998)
Rickettsia prowazekii Madrid E	1,111	834	http://www.uu.se	University of Uppsala	Andersson et al. (1998)
H. pylori J99	1,643	1,495	http://www.genomecorp.com	Astra and Genome Therapeutics	Alm et al. (1999)
Chlamydia pneumoniae CWL029	1,230	1,052	http://www.berkeley.edu	Berkeley University; Incyte Pharmaceuticals	Kalman et al. (1999)
Thermotoga maritima MSB8	1,860	1,877	http://www.tigr.org	TIGR	Nelson et al. (1999)
Methanococcus jannaschii DSM 2661	1,664	1,750	http://www.tigr.org	TIGR; UIUC	Bult et al. (1996)
Methanobacterium thermoautotrophicum DH	1,751	1,918	http://www.genomecorp.com	Genome Therapeutics; Ohio State University	Smith et al. (1997)
Archaeoglobus fulgidus DSM4304	2,178	2,493	http://www.tigr.org	TIGR; UIUC	Klenk et al. (1997)
Pyrococcus horikoshii OT3	1,738	1,979	http://www.miti.go.jp	MITI; University of Tokyo	Kawarabayasi et al. (1998)
Aeropyrum pernix K1	1,669	2,620	http://www.bio.nite.go.jp	Biotechnology Center	Kawarabayasi et al. (1998)
Pyrococcus abyssi GE5	1,765	1,765	http://www.genoscope.cns.fr	Genoscope	
Saccharomyces cerevisiae S288C	12,069	6,294	http://geta.life.uiuc.edu/~nikos/Yeast.html	International collaboration	Mewes et al. (1997)
Caenorhabditis elegans	~97,000	19,099	http://www.genetics.wustl.edu	Washington University; Sanger Center	Anonymous (1998)

ECDC, *E. coli* database collection; *MITI*, Ministry of International Trade and Industry; *ORF*, open reading frame; *TIGR*, The Institute for Genomics Research; *UC*, University of California; *UIUC*, University of Illinois at Urbana-Champaign; *URL*, universal-resource locator

Table 12.2. List of public sequence databases

Name of database	URL	Description of database
Swiss-Prot	http://www.expasy.ch	Protein database
TrEMBL	http://www.expasy.ch	Protein database
PIR	http://pir.georgetown.edu/pirwww/pirhome.html	Protein database
OWL	http://www.leeds.ac.uk/bmb/owl/owl.html	Protein database
GDB	http://www.gdb.org	Genome database
Washington University	http://genome.wustl.edu/est/esthmpg.html	Expressed-sequence tag database
PDB	http://www.rcsb.org/pdb	Protein three-dimensional structures
DDBJ	http://www.ddbj.nig.ac.jp	Nucleotide database
EMBL	http://www.embl-heidelberg.de/Services/index.html	Nucleotide database
NCBI	http://www.ncbi.nlm.nih.gov	Nucleotide database
EcoCyc	http://ecocyc.pangeasystems.com/ecocyc/ecocyc.html	*Escherichia coli* genes and metabolism

DDBJ, DNA Data Bank of Japan; *EMBL*, European Molecular-Biology Laboratory; *GDB*, genome database; *NCBI*, National Center for Biotechnology Information; *PDB*, protein data bank; *URL*, universal-resource locator

– Disease(s) associated with deficiencies(s) of the protein
– Sequence conflicts, variants, etc.

In Swiss-Prot, annotation is mainly found in the comment lines (CC), the feature table (FT) and the keyword lines (KW). Swiss-Prot also tries to minimise the redundancy of sequence data.

12.2.2 TrEMBL

TrEMBL is a supplement of Swiss-Prot (Bairoch and Apweiler 1999) that contains the translations of all European Molecular-Biology Laboratory (EMBL) nucleotide sequence entries not yet entered into Swiss-Prot.

12.2.3 GenBank

GenBank is the National Institutes of Health genetic-sequence database, an annotated collection of all publicly available DNA sequences (Benson et al. 1999). There were approximately 3.4 billion bases in 4.61 million sequence records as of August 1999. A new release is made every 2 months. GenBank is part of the International Nucleotide-Sequence Database Collaboration, which is comprised of the DNA Data Bank of Japan, the EMBL and GenBank at the National Center for Biotechnology Information. These three organisations exchange data on a daily basis.

12.2.4 PIR

The PIR International Protein Sequence Database (Barker et al. 1998) contains information concerning all naturally occurring, wild-type proteins whose primary structures (the sequences) are known. The goal of the database project is to provide a comprehensive, non-redundant database organised by homology and taxonomy. In addition to sequence data, the database contains information concerning: the name and classification of the protein and the organism in which it naturally occurs; references to the primary literature, including information on the sequence determination; the function and general characteristics of the protein; and sites and regions of biological interest within the sequence.

12.2.5 OWL

The OWL sequence database (Bleasby et al. 1994) is a composite, non-redundant database assembled from a number of primary sources, including translations of nucleic-acid sequences. The highest priority is accorded to the Swiss-Prot databank, with the addition of sequences extracted from the National Biomedical Research Foundation/PIR and the Brookhaven Protein Data Bank (PDB) three-dimensional structural database (Sussman et al. 1998). Redundancy is avoided by comparison of sequences, with the elimination of exact duplicates and of sequences that differ only trivially. The data entries include references and other textual information, including cross-references to the PDB three-dimensional structure and PRINTS databases.

12.2.6 EST Databases

ESTs are produced for the public at the University of Washington (Eckman et al. 1998). The only data associated with the ESTs arethe sources of the tissue from which they originated. There are also several private companies that are producing EST databases, such as Celera Genomic Corporation (http://www.celera.com) and Incyte (http://www.incyte.com).

12.3 Two-Dimensional Gel Protein Databases

The development of sensitive and rapid tools for protein identification has stimulated the creation of several comprehensive 2D-gel protein databases during the past few years. In this section, we will describe the publicly available databases. Proteomics databases are being interfaced with the DNA-mapping and genome-sequencing approaches and offer opportunities for the study of protein and gene expression in health and disease. In addition to the plain sequence information one obtains from DNA sequencing, the following data can be obtained in proteomics experiments: the relative abundance of the proteins, PTMs, subcellular localisation, interactions with other proteins and induction by certain stimuli. Proteomics is one of several technologies that can be applied to the field of so-called functional genomics. Other techniques are complementary-DNA arrays, messenger-RNA arrays, phage-display antibody libraries and transgenic animals. An ideal database would link all this information so that one could come closer to understanding protein function. A summary of proteome databases for different cell lines and tissues is given in Table 12.3.

12.3.1 Swiss 2D Polyacrylamide-Gel Electrophoresis

The Swiss 2D polyacrylamide-gel electrophoresis (PAGE; Hoogland et al. 1999) database of proteins identified on 2D-PAGE was built according to the five rules for the creation of a centralised 2D-PAGE database that were proposed by Ron Appel from Geneva. These rules are:

1. Individual entries in the database must be remotely accessible by keyword search. Other query methods are possible but not required (full text searches, for example).
2. The database must be linked to other databases through active hypertext cross-references; i.e. via a simple mouse click on a cross-reference, the user is automatically connected to the corresponding WWW site, and the cross-referenced document is then retrieved and displayed. This simple mechanism links all related databases and combines them into one large virtual database. Database entries must have such a cross-reference, at least to the main index (see rule 3).
3. In addition to individually searchable databases, a main index that provides a means of querying all databases through one unique entry point has to be supplied. Bi-directional cross-references must exist between the main index and the other databases. Currently, the main index is the Swiss-Prot database.
4. Individual protein entries must be accessible via clickable images. That is, 2D images must be provided on the WWW server and, as a response to a mouse click on any identified spot on the image, the user must obtain the

Table 12.3. List of proteome databases for different cell lines and tissues

Biological material	Web location (URL)	Organisation
Liver, plasma, HepG2, HepG2SP, RBC, lymphoma, CSF, macrophage CL, erythroleukaemia CL, platelet, yeast, *Escherichia coli*, colorectal, kidney, muscle, macrophage-like CL, pancreatic islets, epididymus, dictyostelium	http://www.expasy.ch	ExPASy Swiss 2D-PAGE
Mouse liver, human breast cell lines, pyrococcus	http://www.anl.gov/CMB/PMG	Argonne Protein-Mapping Group
Human primary keratinocytes, epithelial, hematopoietic, mesenchymal, hematopoietic, tumors, urothelium, amnion fluid, serum, urine, proteasomes, ribosomes, phosphorylations; mouse epithelial, newborn (ear, heart, liver, lung)	http://biobase.dk/cgi-bin/celis	Danish Centre for Human-Genome Research
Human colorectal CL, placental lysosomes	http://www.ludwig.edu.au/jpsl/jpslhome.html	Joint Protein-Structure Laboratory
A375 melanoma CL	http://rafael.ucsf.edu/2DPAGEhome.html	UCSF 2D-PAGE
E. coli	http://pcsf.brcf.med.umich.edu/eco2dbase	ECO2DBASE (in NCBI repository)
Yeast	http://www.proteome.com	PROTEOME Inc. (YPD)
	http://www.ibgc.u-bordeaux2.fr/YPM	Yeast 2D-gel database, Bordeaux, France
	http://yeast-2dpage-gmm.gu.se	Yeast 2D-PAGE, Göteborg, Sweden
Yeast, REF52, mouse embryo	http://siva.cshl.org/index.html	Quest Protein Database Center
Human, rat and mouse heart	http://www.harefield.nthames.nhs.uk	HSC 2D-PAGE, Harefield Hospital
Human heart	http://www.chemie.fu-berlin.de/user/pleiss	HEART 2D-PAGE, German Heart Institute, Berlin, Germany
Human heart	http://www.mdc-berlin.de/~emu/heart	HP 2D-PAGE, MDC, Berlin, Germany
Rat neuron	http://sunspot.bioc.ac.uk/NEURON.html	Cambridge 2D-PAGE
Embryonic stem cells	http://www.ed.ac.uk/~nh/2DPAGE.html	Immunobiology, University of Edinburgh
Rat, mouse, human liver, corn, wheat	http://www.lsbc.com	Large-Scale Biology Corp
Maize	http://moulon.moulon.inra.fr/imgd	Maize Genome Database, INRA
Drosophila melanogaster	http://tyr.cmb.ki.se	Karolinska Institute

Table 12.3. *Continued*

Biological material	Web location (URL)	Organisation
Bacillus subtilis	http://pc13mi.biologie.uni-greifswald.de	University of Greifswald
Plasma, CSF; urine	http://www-lecb.ncifcrf.gov/PDD	NIMH-NCI PDD
Phosphoprotein, prostate, phosphoprotein, breast cancer drug screen, FAS (plasma), cadmium toxicity (urine), leukaemia	http://www-lecb.ncifcrf.gov/ips-databases.html	IPS/LECB, NCI/FCRDC
Rat liver, kidney, serum, cerebrum; human liver, serum; bovine testis	http://iupucbio1.iupui.edu/frankw/molan.htm	Molecular-Anatomy Laboratory, Indiana University, Purdue University
Human inner ear	http://oto.wustl.edu/thc/innerear2d.htm	Washington University Inner-Ear Protein Database
Mouse brain cerebellum, cortex, hippocammpus, striatum; *Arabidopsis thaliana* callus, leaf, seed, stem; *Oryza sativa* (rice) leaf, callus, germ, root, seed, stem	http://www.rs.noda.sut.ac.jp/~kamom/2de/2d.html	Research Institute for Biological Science, Science University, Tokyo, Japan
Drosophila melanogaster	http://try.cmb.ki.se	*Drosophila melanogaster* at the Karolinska Institute
Cyano2Dbase; *Synechocystis* sp. PCC6803	http://www.kazusa.or.jp/tech/sazuka/cyano/proteome.html	Protein Project for Cyanobacteria
Haemophilus influenzae and *Neisseria meningitidis*	http://www.abdn.ac.uk/~mmb023/2dhome.htm	2D-PAGE Aberdeen
Age-related proteome mapping of TIG-3	http://www.tmig.or.jp/2D/2D_Home.html	TMIG 2D-PAGE
Human leukaemia cell lines	http://www.univ-paris13.fr/biochemistry/Biochimie/biochimie.htm	Laboratorie de Biochimie et Technologie des Proteines, Bobigny, France
Mycobacterium tuberculosis, vaccine strain *Mycobacterium bovis* BCG	http://www.mpiib-berlin.mpg.de/2D-PAGE	Max-Planck-Institut für Infektionsbiologie

BCG, Bacille Calmette-Guerin; *CL*, cell line; *CSF*, cerebrospinal fluid; *FAS*, fatty-acid synthase; *FCRDC*, Frederick Cancer Research and Development Center; *HP*, high-performance; *INRA*, Institut National de la Recherche Agronomique; *IPS*, Image-Processing Section; *LECB*, Laboratory of Experimental and Computational Biology; *MDC*, Max Delbrück Center; *NCBI*, National Center for Biotechnology Information; *NCI*, National Cancer Institute; *NIMH*, National Institutes for Mental Health; *PAGE*, polyacrylamide-gel electrophoresis; *PDD*, protein-disease database; *RBC*, red blood cells; *TMIG*, Tokyo Metropolitan Institute of Gerontology; *UCSF*, University of California at San Francisco; *URL*, universal-resource locator; *YPD*, yeast-proteome database

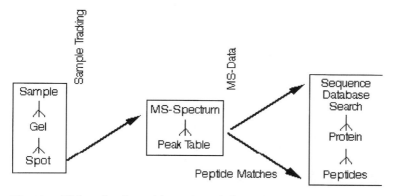

Fig. 12.1. High-molecular-weight region of the *Escherichia coli* proteome map on Expasy, showing the spots that have been identified. A clickable map that links the gel spot directly to the SwissProt entry that it represents is available

database entry for the corresponding protein. This method allows a user to easily identify proteins on a 2D image.

5. 2D analysis software that has been designed for use with central databases must be able to directly access individual entries in any central 2D database. For example, when displaying a 2D reference map with a 2D computer program, the user must be able to select a spot and remotely obtain the corresponding entry from the given database.

The Swiss 2D-PAGE database of proteins identified on 2D-PAGE was built according to these rules. Release 10.0 (June 1999) and updates as recent as 8 September 1999 (which contains 592 entries in 23 reference maps from human, mouse, *Saccharomyces cerevisiae*, *Escherichia coli* and *Dictyostelium discoideum*). There are several ways to access data in this database:

- By description line (DE) or by identity
- By accession number (AC lines)
- By clicking on a spot: select a 2D-PAGE reference map, click on a spot, and get the corresponding information from the Swiss 2D-PAGE database (one of the summary maps for *E. coli* is shown in Fig. 12.1).
- By author (RA lines)
- By full-text search
- Simple repeated sequences, searched for in SWISS 2D-PAGE using the sequence-retrieval system
- Retrieve all the protein entries identified on a given reference map in a table
- Compute-estimated locations on reference maps for a user-entered sequence

The Swiss 2D-PAGE server also provides several tools for proteomics application (a summary of other tools available on the web for protein identification is given in Table 12.4):

Table 12.4. Tools for protein identification available on the Worldwide Web

Name	URL	Description
Protein Prospector	http://prospector.ucsf.edu http://falcon.ludwig.ucl.ac.uk/mshome3.2.htm	A variety of tools from UCSF (MS-Fit, MS-Tag, MS-Digest, etc.) for mining sequence databases in conjunction with mass spectrometry experiments
PeptideSearch	http://www.mann.embl-heidelberg.de/Services/PeptideSearch/PeptideSearchIntro.html	Peptide mass-fingerprint tool from EMBL, Heidelberg, Germany
MOWSE	http://www.seqnet.dl.ac.uk/Bioinformatics/Webapp/mowse	Peptide mass search tool from Daresbury Laboratory
Mascot	http://www.matrixscience.com/cgi/index.pl?page =/search_form_select.html	Peptide mass-fingerprint, sequence-query and MS/MS ion searches from Matrix Science Ltd, London, UK
CombSearch	http://cuiwww.unige.ch/~hammerl4/combsearch	An experimental unified interface that queries several protein-identification tools accessible on the web
MassSearch	http://cbrg.inf.ethz.ch/subsection3_1_3.html	Peptide mass-fingerprinting tool from ETH, Zürich, Switzerland
Prowl	http://www.proteometrics.com	Database search tools that enable swift analysis of protein mass spectrometry data for protein identification and characterisation

EMBL, European Molecular-Biology Laboratory; *ETH*, Eidgenössische Technische Hochschule; *MS/MS*, tandem mass spectrometry; *UCSF*, University of California at San Francisco; *URL*, universal-resource locator

AACompIdent. Identify a protein by its amino acid composition.

AACompSim. Compare the amino acid composition of a Swiss-Prot entry with all other entries.

MultiIdent. Identify proteins with specific pI, molecular weight (MW), amino acid composition, sequence-tag and peptide-mass fingerprinting data.

PeptIdent. Identify proteins with specific peptide-mass fingerprinting data, pI and MWs. Experimentally measured, user-specified peptide masses are compared with the theoretical peptides calculated for all proteins in Swiss-Prot, making extensive use of database annotations.

TagIdent. Identify proteins with specific pI, MW and sequence tags, or generate a list of proteins that have a pI and MW close to a given values.

FindMod. Predict potential protein PTMs and potential single-amino-acid substitutions in peptides. Experimentally measured peptide masses are compared with those of theoretical peptides calculated from a specified Swiss-Prot entry or from a user-entered sequence, and mass differences are used to better characterise the protein of interest.

PeptideMass. Calculate masses of peptides and their PTMs for a Swiss-Prot or TrEMBL entry or for a user sequence.

Compute pI/MW. Compute the theoretical pI and MW from a Swiss-Prot or TrEMBL entry or for a user sequence.

12.3.2 Danish Centre for Human Genome Research 2D-PAGE Databases

The Danish Centre for Human Genome Research's 2D-PAGE databases (Celis et al. 1996) at the University of Aarhus contain data on proteins identified on various reference maps from humans and mice. Protein names and information concerning specific protein spots can be displayed by clicking on the image of interest. Searches by protein name, keywords, MW and pI or organelle or cellular components are possible. Protein files contain extensive links to other databases [Medline, GenBank, Swiss-Prot, PIR, PDB, Online Mendelian Inheritance in Man (OMIM), UniGene, GeneCards, etc.] or web sites. Table 12.5 gives an overview of the Aarhus database.

The databases from Aarhus can be searched in the following ways:

- Search by protein name (using the entire name or part of it).
- Find proteins whose annotation contains the text you specify. If you want to narrow the search, use the AND operator between keywords, e.g. nuclear AND phosphorylated. If you want to expand the search, use the OR operator, e.g. phosphorylated OR glycosylated. If you enter multiple words, they will be treated as a text string, for example, plasma membrane.
- Find a protein by entering its sample spot (SSP) number.
- Find proteins that migrate between ranges of apparent MWs and pIs.

Table 12.5. Contents of the Aarhus proteomics database

Database	Number of proteins catalogued	Number of proteins identified
Human keratinocyte IEF database	2315	865
Human keratinocyte NEPHGE database	956	372
Keratinocyte proteins present in the medium IEF database	358	59
Transitional cell carcinomas IEF database	2171	442
Transitional cell carcinomas NEPHGE database	990	144
Bladder squamous-cell carcinomas IEF database	1578	308
Urine IEF database	459	197
Human MRC-5 fibroblasts IEF database	1440	262
Human MRC-5 fibroblasts NEPHGE database	580	84

IEF, isoelectric focusing; *NEPHGE*, non-equilibrium gradient electrophoresis

- By clicking on an organelle or cellular component, you can obtain a list of proteins in that entry. You can also display their position in the full image. Thereafter, you can view the file for a given protein.

12.3.3 Argonne Protein-Mapping Group Server

The Protein-Mapping Group at Argonne National Laboratory (Williams et al. 1998) has used 2D electrophoresis (2DE) to detect protein changes that occur in the livers of mice exposed to toxic chemicals (peroxisome proliferators) and the offspring of mice exposed to toxic chemicals (*N*-ethyl-*N*-nitrosourea) or ionising radiation (g-rays or neutrons). The mouse-liver master pattern shows the proteins that have been identified thus far. By clicking on any of the high-lighted protein spots, the available identifications can be seen. This centralised WWW database of 2DE patterns currently includes maps of proteins from mouse-liver proteins, human breast-cell proteins and the hyper-thermophilic organism *Pyrococcus furiosus*. The 2DE maps include hyperlinks to textual descriptions of individual protein spots.

12.3.4 SIENA 2D-PAGE

The following maps are available at SIENA 2D-PAGE: human (breast, ductal carcinoma), *Chlamydia trachomatis* (Sanchez-Campillo et al. 1999) and *Caenorhabditis elegans*. This database was created with the tools provided by the Swiss 2D-PAGE database. Approximately 100 different proteins are identified.

12.3.5 HEART 2D-PAGE

HEART 2D-PAGE (Pleissner et al. 1996, 1997a,b) at the German Heart Institute in Berlin is a centralised 2D database of myocardial proteins, identified on 2D-PAGE maps in collaboration with the Wittmann Institute and the Berlin Technical University. There are sets of gels identifying (heart) chamber-specific proteins (right atrium, left ventricle, etc.) and disease-specific proteins, such as those involved in dilated cardiomyopathy.

12.3.6 HSC 2D-PAGE

HSC 2D-PAGE at the Heart Science Centre, Harefield Hospital, UK, provides similar centralised 2D-PAGE databases (Dunn et al. 1997; Evans et al. 1997). This is now a major world site for 2D-gel electrophoresis and is home to the British Society for Electrophoresis' web pages. HSC 2D-PAGE is an advanced interface for accessing protein databases related to heart disease and currently includes databases of proteins from human, dog and rat ventricular tissue and a human endothelial cell line.

12.3.7 National Institutes for Mental Health/National Cancer Institute Protein-Disease Database and the National Cancer Institute/Frederick Cancer Research and Development Center Laboratory of Mathematical Biology Image Processing

National Institutes for Mental Health/National Cancer Institute (NIMH/ NCI) Protein-Disease Database (PDD; Lemkin et al. 1995a,b, 1996; Merril et al. 1995) at the NIMH is part of the project for correlating diseases with proteins observable in serum, cerebrospinal fluid, urine and other common body fluids. Data is being collected and entered into the PDD by the NIMH and others. It may be searched in a variety of ways, including by looking for normalised concentration changes of proteins (for disease states compared with normal states) and other, more complex queries. The PDD may be accessed in order to: (1) find proteins by clicking on spots in a reference 2DE-gel map image, or (2) query the relational database using forms to specify the search query. Query results can then be shown as tables, graphs, 2DE-gel map images or hypertext references to other WWW databases.

NCI/Frederick Cancer Research and Development Center Laboratory of Mathematical Biology (FCRDC LMMB) Image Processing (Lemkin 1997) is the image-processing section of the LMMB, NCI/FCRDC, (Frederick, MD, USA). This web site offers PDD, a centralised 2DE database that correlates proteins with disease.

12.3.8 Yeast-Proteome Database

The Yeast-Proteome Database (YPD; Hodges et al. 1999) is a model for the organisation and presentation of comprehensive protein information. Based on a detailed survey of the scientific literature for the yeast *Saccharomyces cerevisiae*, YPD contains numerous annotations derived from the review of 8500 research publications.

12.3.9 MitoDat

MitoDat (Lemkin et al. 1996) is the mitochondrion database. This partially centralised database specialises in nuclear genes specifying the enzymes, structural proteins and other proteins (many still not identified) involved in mitochondrial biogenesis and function. MitoDat highlights predominantly human nuclear-encoded mitochondrial proteins, but proteins from other animals are included, as are proteins currently known only from yeast and other fungal mitochondria, and plant mitochondria. The database consolidates information from various biological databases, e.g. GenBank, Swiss-Prot, Genome Database and OMIM. The mitochondrion is implicated in many human diseases, so the database should be a valuable tool. MitoDat can be searched for key phrases using Boolean logic operators (AND, OR, etc.). Searches can be restricted to organelle compartments, such as the outer membrane, inter-membrane space, inner membrane and matrix. The database can be downloaded to a local hard disc.

12.3.10 Large-Scale Biology

The Large-Scale Biology (LSB) home page of the LSB company has 2DE-gel maps on-line for standard rat, mouse and human liver (Coomassie Blue-stained 2D patterns), human plasma (Anderson and Anderson 1991), corn and wheat proteins, and 2D non-equilibrium gradient electrophoresis (NEPHGE) of rabbit psoas muscle.

12.3.11 UCSF 2D-PAGE

UCSF 2D-PAGE (A375 cell line) at the University of California, San Francisco (UCSF) presents information on proteins identified after isolation by 2D PAGE (Clauser et al. 1995). The proteins have been identified through extensive use of MS-based peptide sequencing. The work was performed by Dr. Lois B. Epstein's Tumour Immunology and Interferon Research Laboratory at the Cancer Research Institute and the Department of Pediatrics at UCSF and has led to the identification of proteins present in malignant human cell lines.

Of particular interest are proteins that are either induced or suppressed by interferon-g (IFN-g) and/or tumour necrosis factor (TNF), because the identified proteins may mediate the anti-tumour or immunomodulatory functions of the two cytokines. The A375 Melanoma Database contains details about the A375 cell line, a table of the identified melanoma proteins (with links to Swiss-Prot and GenBank entries) and 2DE-gel images of IFN-g/TNF-treated cells and controls.

12.3.12 ECO2DBASE

ECO2DBASE (VanBogelen et al. 1992) is a file-transfer-protocol server for a partially centralised database relating *E. coli* genes to expression products.

12.3.13 Embryonic Stem Cells

The Embryonal Stem Cells database at the University of Edinburgh Immuno-biology Department enables one to obtain molecular insights into cell differentiation, commitment and stem-cell self-renewal. One approach is to study the molecular control of haematopoiesis via a high-resolution map of protein expression at different time points of differentiation and inducing-agent (retinoids) treatment using 2D-PAGE maps.

12.3.14 Human Colon-Carcinoma Protein Database

The Human Colon-Carcinoma Protein Database (Ji et al. 1997; Chataway et al. 1998) is maintained by the Protein-Structure Laboratory in the Ludwig Institute at the University of Melbourne. Their server features updated synthetic images of human colon-carcinoma cell line 2D-gel protein-distribution patterns. Usefully, the site also includes current methods used by the lab for protein extraction from polyacrylamide gel matrices; the site also includes the structural-characterisation techniques.

12.3.15 Yeast 2D-PAGE

Yeast 2D-PAGE (Lundberg Laboratory, University of Goteborg) is a 2D-PAGE protein database containing information about 2D-PAGE-resolved proteins from a number of species of yeast (Norbeck and Blomberg 1997); however, the main focus is on *Saccharomyces cerevisiae*. The database displays information about the identities and expression profiles of resolved proteins.

12.3.16 Haemophilus 2DE Protein Database

The data comprising the 2DE protein database (Cash et al. 1997) can be accessed by a number of different methods, as described below:

- Clickable spot maps for the *H. influenzae* strains HI-64443 and Rd strain are available.
- Enlarged regions of the HI-64443 reference gel are available.
- Displays of spot data from the HI-64443 reference gel are available as a text file.
- Details of the origins of the *Haemophilus* isolates used in the construction of the database are available.
- Representative 2DE protein profiles for the *Haemophilus* isolates are available.

12.4 Proteomics-Database Design

Proteome databases are essential components of any implementation of proteomics technology. The amount of data generated in such experiments is so large that conventional methods of experiment documentation are not feasible. Therefore, good database design is essential to allow the data that is being generated to be examined via high-level data exploration ("data mining") rather than only via the simple analysis that was originally intended. Proteomics databases should reflect the characteristics of the method: the ability to achieve unambiguous, reproducible, quantitative protein separation and the ability to connect this information with protein (gene) identification. The database should contain enough information to reproduce the separation experiment or to recreate the steps leading to the identification of the protein. It should also provide enough information about the sample to arrange sets of samples into useful groups for querying. Few of the current databases fulfil these requirements. In most of the databases, identification information is concerned solely with the location of protein spots on standard 2D-PAGE gels.

Most of the publicly available proteomics databases are accessible via the WWW; we regard this as the only way to publish proteomics-database information that consists of large maps of proteins identified in a particular sample. However, there are two disadvantages with this method of publication: the published data is usually not peer-reviewed, and there are few ways to perform *post factum* analysis of the data. The researcher querying such a web database usually has to take protein identifications for granted; he cannot re-analyse the database with tools other than those the remote server provides (and these are seldom more than simple text-index queries). If the user wants to compare his own data sets with the remote database, he has to reproduce the gel system of the authors and query the information via gel comparison or via direct sequence-database identifying-string comparisons. This situation seems quite unsatisfactory to us. We therefore propose a somewhat different proteomics database: a proteomics database that offers access to the primary data used to derive identification information. Our group has implemented such a system and has gained more than a year of experience with it. However, as database design is a dynamic process, we are only starting to utilise the full power of the design. Therefore, we will refer to features implemented in the Roche Proteomics database and to features that are desirable or are being implemented.

12.4.1 The Database Schema

12.4.1.1 General Remarks

Database systems can be implemented in many vastly different fashions. We will abstract our design schema from the concrete implementation available at Roche. The computing resources, goals and tastes of the implementers may vary from place to place and may develop with time. The Roche database underwent a number of design states: it went from being a collection of static files on a Web server over a shared indexed sequential-access-method database to a multi-user, multi-server, three-tiered Structured-Query Language-based relational database. Despite the large differences in the implementation, the general design schema has been the same from the start:

- A division of the database into three parts (sample tracking, MS-data representation and a cache of identification information)
- A division of the editing interface and the browsing interface
- Hypertext-transfer protocol (HTTP)-based in/out interfaces [Hypertext Markup Language (HTML), Extensible Markup Language (XML)]

The general scheme of the proteomics database is pictured in Fig. 12.2, which should be referred to throughout this chapter.

12.4.1.2 Sample and Gel Tracking

A proteomics database must provide a reasonable amount of information about the biological samples that it contains. Without this information, the user cannot relate the biological objects in the database to each other and would not need to build the database in the first place. However, it should not replace a full labour-management information system and manage a wealth of information unrelated to the goals of a proteomics database. In our view, sample tracking should only be extensive enough to perform the proteomics goals well; therefore, the proteomics database should provide pointers to a

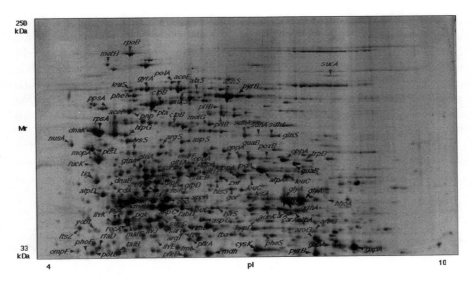

Fig. 12.2. Information flow and organisation in a proteomics database

laboratory-management system while extracting and caching the essential information. In practice, we established the practice of standard experimental protocols and user-defined dictionaries. Thus, the database contains tables with which to classify the protocols used for sample preparation, generate the gels, perform mass-spectrometric data acquisition, stain the gels, etc. The user can query lists of these protocols and can find resources describing the meaning of the entries. However, the user is not allowed to change a protocol once it is established, nor is he allowed to add private protocol descriptions. This is obviously sufficient for a number of stable lab procedures, such as sample preparation and gel casting. However, there are several variable entries when tracking the sample. These entries form a sample-description dictionary and must be allowed to grow with the user's needs. There are various approaches for building these dictionaries, and we cite only the simplest way: there are established conventions for the addition of new entries to the dictionary. These conventions can be enforced either by the database itself or by the user community. Items tracked with such dictionaries include, e.g. the sample source (the biologist who provided us with the sample for analysis), the organism, the organ, the sub-cellular fraction, the pathology, etc.

Experience has shown that these dictionaries are extremely valuable for linking the database with the outside world and for looking at the data outside the original experimental context. Therefore, one should not underestimate the importance of correctly characterising the samples and not allowing sample entries with preliminary or wrong entries. In our group, we have a policy of accepting samples only with a complete description of the relevant terms. However, if information can be selected by an outside browser using the dictionary categories, it follows that the dictionary categories themselves should be standardised throughout the laboratory or, for public-domain databases, throughout the user community. We have found this to be a major issue, as there are few systematic conventions regarding how to categorise, e.g. the pathology of a clinical sample or the strain of a genetically modified organism. It will be important to create useful, context-sensitive, hierarchical dictionaries of biological terms that are as qualified as the biological classification of the species of an organism.

It should be noted that a sample in the Roche proteomics database is a rather specific term that is not fully equivalent to the notion of the sample that a biologist usually has. We consider samples to be entities that can be subjected to 2D-PAGE analysis. Thus, we consider, e.g. fractions of a purification process to be different samples in the context of our database, even though they might be derived from the same biological specimen. One could consider splitting this definition into the specimen and the actual 2D-gel sample, but we do not see a direct advantage of introducing this additional level of information. However, this may change as our research evolves towards a more three-dimensional analysis model where each specimen is subjected to a number of orthogonal sub-fractionations.

12.4.1.3 Spot Location and Quantification Data

It is not disputed that an important task of a proteomics database is that it be a repository of gel images, spot locations and their annotations. If the 2D-PAGE running conditions are chosen and standardised carefully, this kind of information is also of value for cross-referencing gel images and spot positions and conferring (1) protein identification information through the x–y coordinates of protein spots on the gel and (2) quantification data through the dye-binding data at a given gel location. Although we think that a proteomics database should definitely contain this data to facilitate gel comparisons, we have some reservations about the value and usability of this data if it is the only content of the database. Identifying a protein on a 2D-gel through gel locations alone is rather risky due to the fact that the likelihood of a false positive match is more than one in 100 if we take into account the reproducibility of the separation process and the uneven distribution of proteins in MW–pI space. Although reliance on such a poor identification process might be adequate in systems of limited complexity or for highly standardised samples, exploratory experiments with unknown samples or with samples that have high biological variability (clinical samples) require higher levels of accuracy to prevent potentially costly oversights. However, limiting a gel database to gels run under standardised conditions prevents the inclusion of data from other sources. Therefore, we must consider how one can overcome the need for standardised samples and gel-running conditions and still have a chance to reliably identify protein spots on gels that can be compared in differential gene-expression experiments. In our experience, the key to a successful cross-referencing system is to include the data that led to the identification (or that was generated in unsuccessful attempts to identify the protein) in the database.

12.4.1.4 Mass-Spectrometric Data

Given the present state of the art, the prime method for protein identification is mass-spectrometric analysis. In addition, the mass-spectrometric information that is acquired is the most valuable information that can be obtained in a proteomics experiment. Thus, a proteomics database should be a database of mass-spectrometric data. At first, it seems complicated to implement this kind of database. Mass spectra tend to be rather complex, with most of the data due to noise and baseline and only a few selected areas due to peptide peaks. Typically, a full mass spectrum requires 50–120 KB of computer memory, which prohibits the storage of the original data directly in the database. We have implemented algorithms that can extract the relevant information from full spectra and allow the reconstruction of the main features of the spectrum. Our data-analysis software extracts the following information for each identified peptide peak:

- The m/z value for the mono-isotopic peak in the peptide spectrum
- The height of the largest Gaussian and the integral (I) under the whole peptide fitted curve
- The normalised sum of the square of the deviations of the data from the peak fitted curve (c^2)
- The resolution factor of the instrument (R)

These data and a composite number describing the quality of the fit from I, c^2 and R are then entered into the peak table of the database. This is the largest and most important table in our database, and care had to be taken to apply appropriate indexing to it. In our database management system (DBMS), huge performance gains could be achieved using a non-clustered index on the mass column of the table, thus physically ordering the table according to mass values and facilitating common range operations (finding masses between a lower and upper bound). The other columns have clustered indexes, which allow one to quickly join the table to the spot and the MS-tracking part of the database or to order the peaks according to the quality factor, etc. Considering the storage space needed for the indexes, an entry for a single MS peak requires approximately 5 KB of storage space, which allows databases of up to 10^6 spectra on a simple personal-computer-based DBMS. We should point out that the peak table just described can be used for any kind of MS data if the features of the mass spectrum can be described in the (separate) table that is used to identify the constituents of a spectrum. We will describe the utility of the peak table below.

12.4.1.5 Protein Identifications

Most users expect a proteomics database to contain a cache of protein identifications. We want to state that a protein is usually well characterised by its mass spectrometric data alone – one would be able to implement most of the operations of a proteomics database without the protein-identification data. However, for a number of operations, especially those involving knowledge of the matching sequence (like analysis of PTMs), protein identifications are necessary. Protein-identification information does more than record the presumed identity of the protein spot in the database. In fact, the information cached in the proteomics database should allow the reconstruction of all steps in the identification process and give an assessment of the probability that the identification is a false positive (if we consider false positives a major error in the database). At Roche, we record not only the protein names, and their sequence-database identifiers, but also the sequence and location of all matching and non-matching peptides for any peptide-fingerprint/protein-identification match. A database of MS/MS spectra should also record the sequence-tag matches for the identified peptides. Thus, the protein table and the corresponding peptide table are linked to the mass-spectra/peak-table combination using a peptide-match join table – in effect, creating a many-peaks/many-peptides link.

Moreover, the database design should make sure that the sequence matches are not human created but are derived solely by computer analysis. This will exclude subjective influences that can considerably falsify the information in the database. Recently, we described software tools that are able to discriminate between false positives and positives for probabilistic peptide-fingerprint matching. Of course, these tools result in a small number of false negatives but, for a proteomics database, the elimination of possible false positives has a much higher priority than the total number of identifications.

12.4.2 Database Interface

The following section contains some suggestions for actual implementation of a proteomics database. At Roche, we have experimented with a number of implementation designs. While there is no design that will satisfy all the imaginable needs, some of our experiences should be useful for other implementers. We have found that the design and maintenance of a proteomics database is a rather large undertaking and is complicated by the constant and rapid evolution of the supporting technologies (mass spectrometry, dyes for gel staining, etc.). We have created an approach that allows us to cope with changes in a flexible manner. The main design choice was implementation of different levels of interaction with the database via very different tools. Thus, we distinguish between:

Database browsers. These use the database to access biological information, compare spectra and gels, access and search identification information, etc.
Database communicators. These are computer interfaces used to integrate the proteomics database into the corporate environment.
Database editors. These store the sample, gel and spot information and start the spectrum-evaluation batches.
Database agents. These are computer programs that run on a regular schedule to edit and update the database.

12.4.2.1 Browser

The biologist interested in information from a proteomics database is typically a person without a deep understanding of the technicalities of proteomics database design and structure. The end user views the protein mark-up and identification process as a black box that yields identifications for sports with known location and expression levels. The database should do its best to hide the real complexity from these users and provide a simple and logical graphical user interface. Important aspects of the interface are the ability to operate on a subset of samples throughout the user session, to view and compare gel images, and to access sequence-matching information with an emphasis on biological detail (PTMs). Questions summarising large data sets (the occurrence of rare proteins in the current data set, summary information for a whole gel, etc.) are very important for the biological user. The end user should have access to the advanced features of the database (spectra comparisons), with an absolute minimum of free parameters.

All these requirements require a flexible, complex, extensible application framework that propagates, updates and fixes without user intervention and works in a variety of computer environments. We have found the WWW to be an ideal platform for browser applications. The information is presented to the user as a mixture of HTML pages and interactive JAVA applets. The session state is maintained at the server, and communication with the database server is completely hidden by the Web server. There are virtually no requirements (except for the possession of an Intranet connection and a web browser) for the computing environment. The drawback of this approach is that one has to write the application for the lowest common denominator in web technology. This leads to a larger involvement of the web-server/database-server combination than is necessary with state-of-the-art technology. However, for most users, small performance losses are bearable if the user can access the data from their location and with the tools of their choice.

12.4.2.2 Communicators

The second interface to the database is designed not for human users, but for computers. We have found it very useful to expose a computer-parsable public web interface. In this fashion, we handle the interaction of our database with other databases at our company. As technologies like XML evolve, we will see a growth of the need to actually use them to link databases in an intelligent fashion for cross-database queries and links. The advantage of this approach is that the database structure is hidden from the remote user, and the interfaces are well defined without knowledge of the internal workings. Because the creation of the communicator is done at the server site, optimal strategies can be employed to balance the load on the database server. We use communicator interfaces to automatically create sequence-database annotations or to allow automated queries of it.

12.4.2.3 Editor

Whereas reliance on the HTTP protocol has distinct advantages in terms of browsing and communicating the existing content, it is clumsy and slow for interactive editing tasks. These tasks consist of the manual editing or updating of sample, gel and spot entries and automatic entering of the mass-spectrometric data. We have built a client-server application that is capable of doing this by directly interacting with the database. This is a typical fat-client application that transforms and caches all data on the client, allows editing off-line and performs updates during a single transaction. The editors are given to and maintained by a very limited number of people. Thus, the division of the browser and editor interfaces permits efficient access control.

12.4.2.4 Agents

The database is edited by agents that update certain parts of the database at regular intervals. These agents run in the environment of the database administrator and solve editing tasks, such as sequence-database searching. The agent for this search has evolved considerably and is now able to keep the database consistent with the updates of the sequence databases, guaranteeing that MS data without satisfactory identification will be re-searched at 3-month intervals. Thus, for the purposes of the proteomics scientist, the evaluation and interpretation of the MS data is hidden and performed by computer programs without user intervention. This is convenient, gives the researcher more time to consider biological questions and leads to consistency in data interpretation and higher confidence in the results.

12.4.3 Database Tasks

We now describe a number of tasks that are typical for a proteomics database. As of the time of this writing, few databases have facilities enabling them to carry out the suggested analyses. We will show that an MS-data-based proteomics database can perform most of the tasks easily.

12.4.3.1 Mapping of Proteomes

The main task of a proteomics database is the storage of maps of the proteome for the biological objects investigated. This is the task that most of the proteomics databases already do well. However, most of the databases are based on the comparison of spot locations for standard gels or protein identifications. Gel-independent maps using MS-fingerprint data can be drawn.

12.4.3.2 Protein-Expression Analysis

Once a protein can be identified (in the sense that it can be unambiguously located using experimental data), we can use various approaches to quantify its expression. The MS-data-based approach offers the advantage that it makes it easy to group related spectra (and, hence, related proteins) together. However, reliable expression analysis can also be done without a similar protein grouping by comparing the gels on a spot-by-spot basis and adding expression levels for spots, recalling that one must refer to the same protein identification post factum. This is the approach taken by most of the off-line and on-line software tools we know of today.

12.4.3.3 Integrated Genomic Analysis

The integration of proteomics with other kinds of genome analysis is a different task and, to our knowledge, is not solved in the public domain. In fact, we have a task that is just the opposite of the normal proteomics identification tasks. Whereas we are normally asked to compare a fingerprint spectrum or a sequence tag with a database of proteomics sequences, we now have a gene (ideally corresponding to a single protein sequence, but usually corresponding to a number of variants and homologues) that we must compare with a number of spectra. We want to state that it is usually not sufficient to compare the gene names or database identifiers that a genome experiment yields against the protein identifiers cached in the proteomics database. This will typically yield only the most abundant proteins for which identifiers have been established well enough to be compared in genome and protein databases. Once again, the use of MS data as the primary data in the database makes it easy to implement comparisons against such sequences.

12.4.3.4 Protein Modifications

One decisive advantage of proteomics is the ability to draw conclusions regarding PTMs of proteins. Several tools have implemented the search for PTMs. We are most comfortable with variants that identify proteins based on unmodified sequence-database entries (with the addition of possibly the most common modification artefact: methionine oxidation). Inclusion of the modification variants in the primary search increases the target-sequence database size dramatically and increases the chance of encountering false-positive misidentifications. The analysis of PTMs can be done *post factum* when comparing the expected (calculated) spectrum with the one that was experimentally obtained. Large, unexplained spectra peaks can then be scrutinised on the basis of spectra comparisons for MS data sets not identified by normal sequence-database searches; also, the amino acid sequence of database matches can be used to predict PTM sites and search for the corresponding spectrum peaks.

References

Alm RA, Ling LS, Moir DT, King BL, Brown ED, Doig PC, Smith DR, Noonan B, Guild BC, deJonge BL, Carmel G, Tummino PJ, Caruso A, Uria-Nickelsen M, Mills DM, Ives C, Gibson R, Merberg D, Mills SD, Jiang Q, Taylor DE, Vovis GF, Trust TJ (1999) Genomic-sequence comparison of two unrelated isolates of the human gastric pathogen *Helicobacter pylori*. Nature 397:176–180

Anderson NL, Anderson NG (1991) A two-dimensional gel database of human plasma proteins. Electrophoresis 12:883–906

Andersson SG, Zomorodipour A, Andersson JO, Sicheritz-Ponten T, Alsmark UC, Podowski RM, Naslund AK, Eriksson AS, Winkler HH, Kurland CG (1998) The genome sequence of *Rickettsia prowazekii* and the origin of mitochondria. Nature 396:133–140

Anonymous (1998) Genome sequence of the nematode *C. elegans*: a platform for investigating biology. The *C. elegans* sequencing consortium. Science 282:2012–2018

Bairoch A, Apweiler R (1999) The SWISS-PROT protein sequence data bank and its supplement TrEMBL in 1999. Nucleic Acids Res 27:49–54

Barker WC, Garavelli JS, Haft DH, Hunt LT, Marzec CR, Orcutt BC, Srinivasarao GY, Yeh LSL, Ledley RS, Mewes HW, Pfeiffer F, Tsugita, A (1998) The PIR-international protein sequence database. Nucleic Acids Res 26:27–32

Benson DA, Boguski MS, Lipman DJ, Ostell J, Ouellette BF, Rapp BA, Wheeler DL (1999) GenBank. Nucleic Acids Res 27:12–17

Berndt P, Hobohm, U, Langen H (1999) Reliable automatic protein identification from matrix-assisted laser desorption and ionisation mass spectrometric peptide fingerprints. Electrophoresis 20:3521–3526

Blattner FR, Plunkett G III, Bloch CA, Perna NT, Burland V, Riley M, Collado-Vides J, Glasner JD, Rode CK, Mayhew GF, Gregor J, Davis NW, Kirkpatrick HA, Goeden MA, Rose DJ, Mau B, Shao Y (1997) The complete genome sequence of *Escherichia coli* K-12. Science 277:1453–1474

Bleasby AJ, Akrigg D, Attwood TK (1994) OWL–a non-redundant composite protein sequence database. Nucleic Acids Res 22:3574–3577

Bult CJ, White O, Olsen GJ, Zhou L, Fleischmann RD, Sutton GG, Blake JA, FitzGerald LM, Clayton RA, Gocayne JD, Kerlavage AR, Dougherty BA, Tomb JF, Adams MD, Reich CI, Overbeek R, Kirkness EF, Weinstock KG, Merrick JM, Glodek A, Scott JL, Geoghagen NSM, Venter JC (1996) Complete genome sequence of the methanogenic archaeon, *Methanococcus jannaschii*. Science 273:1058–1073

Cash P, Argo E, Langford PR, Kroll JS (1997) Development of a *Haemophilus* two-dimensional protein database. Electrophoresis 18:1472–1482

Celis JE, Gromov P, Ostergaard M, Madsen P, Honore B, Dejgaard K, Olsen E, Vorum H, Kristensen DB, Gromova I, Haunso A, Van Damme J, Puype M, Vandekerckhove J, Rasmussen HH (1996) Human 2-D PAGE databases for proteome analysis in health and disease: http://biobase.dk/cgi-bin/celis. FEBS Lett 398:129–134

Chalfie M (1998) Genome sequencing. The worm revealed. Nature 396:620–621

Chataway TK, Whittle AM, Lewis MD, Bindloss CA, Moritz RL, Simpson RJ, Hopwood JJ, Meikle PJ (1998) Development of a two-dimensional gel electrophoresis database of human lysosomal proteins. Electrophoresis 19:834–846

Clauser KR, Hall SC, Smith DM, Webb JW, Andrews LE, Tran HM, Epstein LB, Burlingame AL (1995) Rapid mass spectrometric peptide sequencing and mass matching for characterization of human melanoma proteins isolated by two-dimensional PAGE. Proc Natl Acad Sci USA 92:5072–5076

Cole ST, Brosch R, Parkhill J, Garnier T, Churcher C, Harris D, Gordon SV, Eiglmeier K, Gas S, Barry CE III, Tekaia F, Badcock K, Basham D, Brown D, Chillingworth T, Connor R, Davies R, Devlin K, Feltwell T, Gentles S, Hamlin N, Holroyd S, Hornsby T, Jagels K, Barrell BG et al. (1998) Deciphering the biology of *Mycobacterium tuberculosis* from the complete genome sequence. Nature 393:537–544

Deckert G, Warren PV, Gaasterland T, Young WG, Lenox AL, Graham DE, Overbeek R, Snead MA, Keller M, Aujay M, Huber R, Feldman RA, Short JM, Olsen GJ, Swanson RV (1998) The complete genome of the hyperthermophilic bacterium *Aquifex aeolicus*. Nature 392:353–358

Dunn MJ, Corbett JM, and Wheeler CH (1997) HSC-2DPAGE and the two-dimensional gel electrophoresis database of dog heart proteins. Electrophoresis 18:2795–2802

Eckman BA, Aaronson JS, Borkowski JA, Bailey WJ, Elliston KO, Williamson AR, Blevins RA (1998) The Merck Gene Index browser: an extensible data integration system for gene finding, gene characterization and EST data mining. Bioinformatics 14:2–13

Evans G, Wheeler CH, Corbett JM, Dunn MJ (1997) Construction of HSC-2DPAGE: a two-dimensional gel electrophoresis database of heart proteins. Electrophoresis 18:471–479

Fleischmann RD, Adams MD, White O, Clayton RA, Kirkness EF, Kerlavage AR, Bult CJ, Tomb JF, Dougherty BA, Merrick JM et al. (1995) Whole-genome random sequencing and assembly of *Haemophilus influenzae* Rd. Science 269:496–512

Fountoulakis M, Juranville JF, Roder D, Evers S, Berndt P, Langen, H (1998a) Reference map of the low molecular mass proteins of *Haemophilus influenzae*. Electrophoresis 19:1819–1827

Fountoulakis M, Takacs B, Langen H (1998b) Two-dimensional map of basic proteins of *Haemophilus influenzae*. Electrophoresis 19:761–766

Fraser CM, Gocayne JD, White O, Adams MD, Clayton RA, Fleischmann RD, Bult CJ, Kerlavage AR, Sutton G, Kelley JM et al (1995) The minimal gene complement of *Mycoplasma genitalium*. Science 270:397–403

Fraser CM, Casjens S, Huang WM, Sutton GG, Clayton R, Lathigra R, White O, Ketchum KA, Dodson R, Hickey EK, Gwinn M, Dougherty B, Tomb JF, Fleischmann RD, Richardson D, Peterson J, Kerlavage AR, Quackenbush J, Salzberg S, Hanson M, van Vugt R, Palmer N, Adams MD, Gocayne J, Venter JC et al (1997) Genomic sequence of a Lyme disease spirochaete, *Borrelia burgdorferi*. Nature 390:580–586

Fraser CM, Norris SJ, Weinstock GM, White O, Sutton GG, Dodson R, Gwinn M, Hickey EK, Clayton R, Ketchum KA, Sodergren E, Hardham JM, McLeod MP, Salzberg S, Peterson J, Khalak H, Richardson D, Howell JK, Chidambaram M, Utterback T, McDonald L, Artiach P, Bowman C, Cotton MD, Venter JC et al. (1998) Complete genome sequence of *Treponema pallidum*, the syphilis spirochete. Science 281:375–388

Henzel WJ, Billeci TM, Stults JT, Wong SC, Grimley C, Watanabe, C (1993) Identifying proteins from two-dimensional gels by molecular mass searching of peptide fragments in protein sequence databases. Proc Natl Acad Sci USA 90:5011–5015

Himmelreich R, Hilbert H, Plagens H, Pirkl E, Li BC, Herrmann, R (1996) Complete sequence analysis of the genome of the bacterium *Mycoplasma pneumoniae*. Nucleic Acids Res 24:4420–4449

Hodges PE, McKee AH, Davis BP, Payne WE, Garrels JI (1999) The Yeast Proteome Database (YPD): a model for the organization and presentation of genome-wide functional data. Nucleic Acids Res 27:69–73

Hoogland C, Sanchez JC, Tonella L, Bairoch A, Hochstrasser DF, Appel RD (1999) The SWISS-2DPAGE database: what has changed during the last year. Nucleic Acids Res 27:289–291

James P, Quadroni M, Carafoli E, Gonnet G (1993) Protein identification by mass profile fingerprinting. Biochem Biophys Res Commun 195:58–64

Ji H, Reid GE, Moritz RL, Eddes JS, Burgess AW, Simpson RJ (1997) A two-dimensional gel database of human colon carcinoma proteins. Electrophoresis 18:605–613

Kalman S, Mitchell W, Marathe R, Lammel C, Fan J, Hyman RW, Olinger L, Grimwood J, Davis RW, Stephens RS (1999) Comparative genomes of *Chlamydia pneumoniae* and *C. trachomatis*. Nat Genet 21:385–389

Kaneko T, Sato S, Kotani H, Tanaka A, Asamizu E, Nakamura Y, Miyajima N, Hirosawa M, Sugiura M, Sasamoto S, Kimura T, Hosouchi T, Matsuno A, Muraki A, Nakazaki N, Naruo K, Okumura S, Shimpo S, Takeuchi C, Wada T, Watanabe A, Yamada M, Yasuda M, Tabata S (1996) Sequence analysis of the genome of the unicellular cyanobacterium *Synechocystis* sp. strain PCC6803. II. Sequence determination of the entire genome and assignment of potential protein-coding regions (supplement). DNA Res 3:185–209

Kawarabayasi Y, Sawada M, Horikawa H, Haikawa Y, Hino Y, Yamamoto S, Sekine M, Baba S, Kosugi H, Hosoyama A, Nagai Y, Sakai M, Ogura K, Otsuka R, Nakazawa H, Takamiya M, Ohfuku Y, Funahashi T, Tanaka T, Kudoh Y, Yamazaki J, Kushida N, Oguchi A, Aoki K, Kikuchi H (1998) Complete sequence and gene organization of the genome of a hyper-thermophilic archaebacterium, *Pyrococcus horikoshii* OT3. DNA Res 5:55–76

Klenk HP, Clayton RA, Tomb JF, White O, Nelson KE, Ketchum KA, Dodson RJ, Gwinn M, Hickey EK, Peterson JD, Richardson DL, Kerlavage AR, Graham DE, Kyrpides NC, Fleischmann RD, Quackenbush J, Lee NH, Sutton GG, Gill S, Kirkness EF, Dougherty BA, McKenney K, Adams MD, Loftus B, Venter JC et al (1997) The complete genome sequence of the hyperthermophilic, sulphate-reducing archaeon *Archaeoglobus fulgidus*. Nature 390:364–370

Klose J (1975) Protein mapping by combined isoelectric focusing and electrophoresis of mouse tissues. A novel approach to testing for induced point mutations in mammals. Humangenetik 26:231–243

Kunst F, Ogasawara N, Moszer I, Albertini AM, Alloni G, Azevedo V, Bertero MG, Bessieres P, Bolotin A, Borchert S, Borriss R, Boursier L, Brans A, Braun M, Brignell SC, Bron S, Brouillet S, Bruschi CV, Caldwell B, Capuano V, Carter NM, Choi SK, Codani JJ, Connerton IF, Danchin A et al. (1997) The complete genome sequence of the gram-positive bacterium *Bacillus subtilis*. Nature 390:249–256

Langen H, Roder D, Juranville JF, Fountoulakis M (1997) Effect of protein application mode and acrylamide concentration on the resolution of protein spots separated by two-dimensional gel electrophoresis. Electrophoresis 18:2085–2090

Lemkin PF (1997) Comparing two-dimensional electrophoretic gel images across the Internet. Electrophoresis 18:461–470

Lemkin PF, Orr GA, Goldstein MP, Creed GJ, Myrick JE, Merril, C. R (1995a) The Protein Disease Database of human body fluids: II. Computer methods and data issues. Appl Theor Electrophor 5:55–72

Lemkin PF, Orr GA, Goldstein MP, Creed J, Whitley E, Myrick JE, Merril CR (1995b) The protein disease database. Electrophoresis 16:1175

Lemkin PF, Chipperfield M, Merril C, Zullo S (1996) A World Wide Web (WWW) server database engine for an organelle database, MitoDat. Electrophoresis 17:566–572

Merril CR, Goldstein MP, Myrick JE, Creed GJ, Lemkin PF (1995) The protein disease database of human body fluids: I. Rationale for the development of this database. Appl Theor Electrophor 5:49–54

Mewes HW, Albermann K, Bahr M, Frishman D, Gleissner A, Hani J, Heumann K, Kleine K, Maierl A, Oliver SG, Pfeiffer F, Zollner A (1997) Overview of the yeast genome. Nature 387:7–65

Molloy MP, Herbert BR, Walsh BJ, Tyler MI, Traini M, Sanchez JC, Hochstrasser DF, Williams KL, Gooley AA (1998) Extraction of membrane proteins by differential solubilization for separation using two-dimensional gel electrophoresis. Electrophoresis 19:837–844

Nelson KE, Clayton RA, Gill SR, Gwinn ML, Dodson RJ, Haft DH, Hickey EK, Peterson JD, Nelson WC, Ketchum KA, McDonald L, Utterback TR, Malek JA, Linher KD, Garrett MM, Stewart AM, Cotton MD, Pratt MS, Phillips CA, Richardson D, Heidelberg J, Sutton GG, Fleischmann RD, Eisen JA, Fraser CM et al (1999) Evidence for lateral gene transfer between Archaea and bacteria from genome sequence of *Thermotoga maritima*. Nature 399:323–329

Norbeck, J, Blomberg A (1997) Two-dimensional electrophoretic separation of yeast proteins using a non-linear wide range (pH 3–10) immobilized pH gradient in the first dimension; reproducibility and evidence for isoelectric focusing of alkaline (pI > 7) proteins. Yeast 13:1519–1534

O'Farrell PH (1975) High resolution two-dimensional electrophoresis of proteins. J Biol Chem 250:4007–4021

Patterson SD, Aebersold R (1995) Mass spectrometric approaches for the identification of gel-separated proteins. Electrophoresis 16:1791–1814

Pleissner KP, Sander S, Oswald H, Regitz-Zagrosek V, Fleck E (1996) The construction of the World Wide Web-accessible myocardial two-dimensional gel electrophoresis protein database "HEART-2DPAGE": a practical approach. Electrophoresis 17:1386–1392

Pleissner KP, Sander S, Oswald H, Regitz-Zagrosek V, Fleck E (1997a) Towards design and comparison of World Wide Web-accessible myocardial two-dimensional gel electrophoresis protein databases. Electrophoresis18:480–483

Pleissner KP, Soding P, Sander S, Oswald H, Neuss M, Regitz-Zagrosek V, Fleck E (1997b) Dilated cardiomyopathy-associated proteins and their presentation in a WWW-accessible two-dimensional gel protein database. Electrophoresis 18:802–808

Sanchez-Campillo M, Bini L, Comanducci M, Raggiaschi R, Marzocchi B, Pallini V, Ratti G (1999) Identification of immunoreactive proteins of *Chlamydia trachomatis* by Western blot analysis of a two-dimensional electrophoresis map with patient sera. Electrophoresis 20:2269–2279

Smith DR, Doucette-Stamm LA, Deloughery C, Lee H, Dubois J, Aldredge T, Bashirzadeh R, Blakely D, Cook R, Gilbert K, Harrison D, Hoang L, Keagle P, Lumm W, Pothier B, Qiu D, Spadafora R, Vicaire R, Wang Y, Wierzbowski J, Gibson R, Jiwani N, Caruso A, Bush D, Reeve JN et al (1997) Complete genome sequence of *Methanobacterium thermoautotrophicum* deltaH: functional analysis and comparative genomics. J Bacteriol 179:7135–7155

Stephens RS, Kalman S, Lammel C, Fan J, Marathe R, Aravind L, Mitchell W, Olinger L, Tatusov RL, Zhao Q, Koonin EV, Davis RW (1998) Genome sequence of an obligate intracellular pathogen of humans: *Chlamydia trachomatis*. Science 282:754–759

Sussman JL, Lin D, Jiang J, Manning NO, Prilusky J, Ritter O, Abola EE (1998) Protein Data Bank (PDB): database of three-dimensional structural information of biological macromolecules. Acta Crystallogr D Biol Crystallogr 54:1078–1084

Tomb JF, White O, Kerlavage AR, Clayton RA, Sutton GG, Fleischmann RD, Ketchum KA, Klenk HP, Gill S, Dougherty BA, Nelson K, Quackenbush J, Zhou L, Kirkness EF, Peterson S, Loftus B, Richardson D, Dodson R, Khalak HG, Glodek A, McKenney K, Fitzegerald LM, Lee N,

Adams MD, Venter JC et al. (1997) The complete genome sequence of the gastric pathogen *Helicobacter pylori*. Nature 388:539–547

VanBogelen RA, Sankar P, Clark RL, Bogan JA, Neidhardt FC (1992) The gene-protein database of *Escherichia coli*: edition 5. Electrophoresis 13:1014–1054

Williams K, Chubb C, Huberman E, Giometti CS (1998) Analysis of differential protein expression in normal and neoplastic human breast epithelial cell lines. Electrophoresis 19:333–343

Yates JR III (1998) Mass spectrometry and the age of the proteome. J Mass Spectrom 33:1–19

13 Quo Vadis

PETER JAMES

13.1 Introduction

The success of the genome projects, when measured by the sheer amount of sequence data that has been generated, is immense. However, the number of genes for which a function can be assigned is rather meagre and, hence, the discipline "functional genomics" has been created to describe the analysis of gene expression and function. The genome of the yeast *Saccharomyces cerevisiae* contains at least 6200 genes. Despite intensive genetic work during recent years, 60% of yeast genes have no assigned function, and half of these encode putative proteins without any homology with known proteins (Goffeau et al. 1996). In order to describe the functions of the yeast genes, a systematic, large-scale approach is being taken using a combination of mutant generation and analysis by transcriptomics, proteomics and metabolomics (Oliver et al. 1998). The exponential growth of sequences in the database will not make it easier to allocate function by homology; indeed many of the new genes being discovered are unique to certain species (especially in bacteria).

Proteome analysis can be subdivided into expression proteomics, which analyses protein expression and modification, and cell-map proteomics, which attempts to define all protein–protein interactions occurring in a cell under given conditions (Fig. 13.1; Blackstock and Weir 1999). Expression proteomics relies heavily on quantitative two-dimensional polyacrylamide-gel electrophoresis (2D-PAGE) to map protein expression in defined cells and to follow the way protein expression changes in response to perturbation (both genetic and environmental modifications). Cell-map proteomics can be performed either via high-throughput genetic screening using two-hybrid systems (Fromont-Racine et al. 1997) or via the isolation and characterisation of protein complexes. Given this biological backdrop, one should ask what role mass spectrometry (MS) can play. Is MS limited to rapid protein identification for expression proteomics, or can its capabilities be extended to cell-map proteomics? In this chapter, recent developments and trends in MS will be highlighted, and emerging, novel applications that indicate the possible roles MS will play in proteomics will be described.

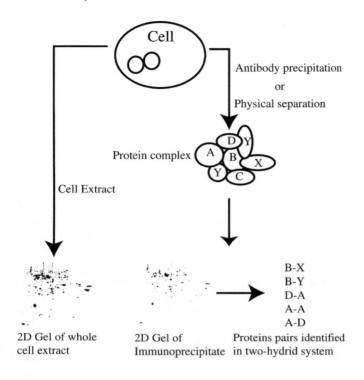

Fig. 13.1. A schematic representation of expression and cell-map proteomics

13.2 Plug-and-Play MS

A deliberately partisan choice was made in the selection of MS instruments for presentation in this book. Matrix-assisted laser desorption ionisation time-of-flight (MALDI-TOF), triple-quadrupole (QQQ) and ion-trap (IT) instruments were chosen, because they have the largest installed user base in protein-chemistry laboratories and service facilities. There is, however, a wide spectrum of instruments that have been and can be constructed. One can divide the available MS components into several groups: ion generation, separation, dissociation and detection. These can be combined as one wishes. The main ion-generation sources are: electrospray ionisation (ESI), MALDI, gas chromatography and electron and chemical ionisation. TOF, ion-cyclotron resonance (ICR), quadrupole IT, quadrupoles (including, hexa- and octopoles etc.) and magnetic sectors are the most common separation units. These can be combined to give hybrid instruments in a variety of geometries (QQQ,

Q-TOF, IT-IT, TOF-TOF, EB-IT etc). A large number of fragmentation methods are available: gas cells allow collisionally activated dissociation; infrared and ultraviolet (UV) lasers can provide photo-fragmentation; the geometry of the instrument can allow ions to collide with a solid surface for surface-induced fragmentation; and the ion containment area can be heated up to cause black-body infrared-radiative dissociation. There are a smaller variety of detectors, such as secondary electron multipliers, Faraday cups, etc. Thus, the number of instruments that can be made by a simple pick-and-mix procedure using the various components mentioned above is very large.

13.3 Recent Advances in MS Instrumentation

13.3.1 Sensitivity and Duty Cycles

For all instruments that use either secondary electron multipliers or array detectors as detection devices, the absolute limit of sensitivity is a single ion. This is due to the cascade-like amplification; the arrival of a single ion triggers the release of tens of electrons, which are accelerated and collide with a wall, causing the release of hundreds of electrons, and so on in cascade fashion. However, QQQ and magnetic-sector mass spectrometers are highly inefficient when one considers the ratio of ions entering the instrument to the number detected. This is a characteristic common to all beam-type instruments that scan a mass range using a small *mass-selective ion-stability* window to allow ions to pass and reach the detector. The ratio of the window width to the total mass range being scanned determines the fraction of time that a given ion can be transmitted during a scan. This is called the "duty cycle" and is less than 1%; i.e. 99% of the ions entering the instrument are lost while a small m/z range is detected. This can be increased using techniques such as single-ion monitoring or the use of an array detector. Recently, a quadrupole/orthogonal-acceleration TOF (Q-TOF; Morris et al. 1996) mass spectrometer that operates with a duty cycle near 50% has been developed. This gives an increase of between one and two orders of magnitude in sensitivity and a similar increase in resolving power, due to the use of the TOF analyser.

In contrast to beam instruments, the quadrupole IT uses *mass-selective ion instability* to obtain a spectrum (McLuckey et al. 1994). Ions are accumulated in the trap until a maximum ion density is reached. The mass spectrum is obtained by scanning the RF ion-trapping frequency, sequentially bringing ions into resonance with the trapping potential. This causes mass-dependent ion ejection from the trap into the detector. For high-sensitivity analyses, such as nanospray, ion-accumulation times can often be 1 s or more, while the mass analysis time is approximately 50 ms. Thus, the duty cycle can be 95% or more. For pulsed instruments, such as MALDI-TOF mass spectrometers, the duty cycle is 100%, because material is only consumed during the laser pulse that initiates measurement. If the efficiency of ion fragmentation can be increased in a controlled and reproducible fashion, MALDI-TOF could find greater acceptance in sequencing applications. However, perhaps the most promising MS technique that could come into widespread use in the future (if it can be manufactured cheaply enough) is Fourier-transform ICR MS (FT-ICR MS).

13.3.2 Miniaturisation

In order to be able to perform high-throughput analyses of ever-smaller amounts of protein, new techniques for sample handling will have to be developed. As one can read in many chapters in this book, the use of nanotechnology to construct micro-machined tools is being actively explored. Many separation techniques, such as high-performance liquid chromatography (HPLC) and capillary zone electrophoresis (CZE), have already been produced in micro-formats, and enzymatic and chemical reactors have been developed in order to perform modifications. The logical extension of this is the construction of micro mass spectrometers; the construction and operation of miniature quadrupole (Tunstall et al. 1998) and IT (Kornienko et al. 1999) mass spectrometers have been described. Another development arising from the silicon-chip technology was the discovery that highly porous silica surfaces can act as matrix-free supports for laser desorption and ionisation (Wei et al. 1999). The silica can be derivatised with C18 reversed-phase material for protein and peptide immobilisation and, because it has such a high UV absorbance, it can act as a matrix on which to trap and transfer the energy from a laser pulse. The main advantages are that there are no matrix ions (so one can easily see low-mass ions, such as Na^+, m/z = 23) and there is very little fragmentation.

13.4 Fourier-Transform Ion-Cyclotron Resonance Mass Spectrometry

13.4.1 The Advantages of FT-ICR MS

13.4.1.1 Extreme Sensitivity and Resolution

FT-ICR is potentially one of the most sensitive and widely applicable of all MS instruments, and interest in it has recently been revived, although it was first described in 1974 by Comisarow and Marshall. The instrument has impressive performance capabilities. A resolving power of over 100,000 has been demonstrated for proteins with a mass of 100 kDa with an accuracy of ±3 Da

(Kelleher et al. 1997). The upper-limit factor is the distribution of low-abundance isotopes, such as ^{13}C, ^{15}N, ^{18}O, etc. This can be alleviated by the use of maximum-entropy-based deconvolution (Zhang et al. 1997) to narrow the isotopic distribution or by cultivating the organism or cell producing the protein in a medium depleted of isotopic ^{13}C and ^{15}N (Marshall et al. 1997).

A resolution of almost one million was achieved with a peptide having a m/z of 650, allowing the separation of the isotopes ^{12}C, ^{13}C, etc. and the complete resolution of the ^{34}S and ^{13}C isotopes, which are separated by only 0.011 Da (Solouki et al. 1997). The mass range is also enormous; for example, a single ionised molecule of coliphage T4 DNA was trapped, and the mass was determined to be 1.1×10^8 amu (Chen et al. 1995). Single ions can be measured although, in general, a few hundred ions should be trapped to give a clear current. Sub-attomole sensitivity (<600,000 molecules) for MS and tandem MS (MS/MS) of intact peptides and proteins with masses as high as 30 kDa has been demonstrated (Valaskovic et al. 1996).

13.4.1.2 Non-Destructive Measurement

The main distinguishing feature of the technique is that ion detection is non-destructive (Guan and Marshall 1997). Unlike beam or quadrupole IT instruments that rely on destructive measurement by causing collisions between ions and detectors, FT-ICR MS uses the non-destructive method of image-current detection. Essentially, ions are injected into the centre of a Penning trap, which is formed by a combination of a static magnetic field and a quadrupolar electrostatic trapping field. The ions can be visualised by applying a radio-frequency (RF) electric field with the same frequency as the ion cyclotron frequency (i.e. producing resonance). The movement in the magnetic field induces an image current in two conductive, directly opposing, parallel electrodes, which act as the detector. A very short, intense electric-field pulse produces broadband electromagnetic radiation over a frequency range that is inversely proportional to the pulse duration. Each of the m/z values is stimulated in the time domain, and the measured response can be converted to the frequency domain (the normal mass spectrum) by Fourier transformation. For a comprehensive account of the theory and practice of FT-ICR, there is an excellent tutorial on the subject by the creator of the instrument (Marshall et al. 1998).

The non-destructive nature of measurement with this instrument has an important consequence: after mass determination, it is theoretically possible to purify the sample by molecular mass selection and to recover it from the instrument for further use. There are many possible applications of this, such as the recovery of molecules from highly complex mixtures for biological-activity measurements. These include: DNA recovery for polymerase chain reaction (PCR), major histocompatibility complex-presented peptide recovery for antigenicity testing, etc. Very recently, the recovery of DNA fragments for

PCR has been demonstrated by Smith's group (Feng et al. 1999) using a technique they call "soft-landing".

13.4.2 The Disadvantages of FT-ICR MS

13.4.2.1 High Vacuum

The main limitations of the technique have been the need to operate at a very high vacuum (10^{-8} Torr), the low duty cycles and the necessity for long data-accumulation times. In order to couple FT-ICR with external atmospheric ionisation techniques, such as ESI, the ion source must be separated from the trap by several stages of differential pumping. Recently, powerful cryo-pumps and the very small aperture inlets used with nanospray techniques have provided the necessary vacuum. Since the ions are generated external to the magnetic trap, they have to be injected through the magnetic-field fringe by ion guides, such as quadrupoles or octopoles (Senko et al. 1997). Another problem is trapping the ions once they enter the cyclotron, especially the highly energetic ions generated by MALDI. The ions can be cooled by pulsing an inert gas into the MS to provide collisional cooling, but this means another time loss of a few seconds in order to remove the gas before measurement.

13.4.2.2 Duty Cycle

The poor duty cycle, especially using on-line separation techniques at atmospheric pressure (HPLC/CZE-ESI), is due mainly to the time required to pump the instrument to operating pressures after ion injection. The data sets generated by FT-ICR are extremely large, because a huge number of data points must be collected in order to take advantage of the resolving power; hence, a second limiting factor is the time required to store the data. The actual measurement times are quite short (a few hundred microseconds). These problems are being solved with the development of new data systems (Senko et al. 1996; Guan and Marshall 1997) and instrument designs (Littlejohn and Ghaderi 1986; Carroll et al. 1996). One exciting development is the accumulation of external generated ions in a device (such as a linear quadrupole trap) prior to injection into the cyclotron (Senko et al. 1997). The injected ions can be measured immediately without the need for a long pumping period, and scan/data accumulation rates of one spectrum per 2 s, compatible with liquid-chromatography and CZE separations, can be achieved. There are no ions lost during this 2 s, because a new set of ions are being accumulated externally again, giving a duty cycle of 100%.

13.5 Future Directions

13.5.1 Bypassing Proteolysis: Fragmenting Whole Proteins in the Mass Spectrometer

13.5.1.1 Fragmentation Methods

Currently, the major limiting factor in biological protein and peptide analysis is sample handling. Enzymatic digestion of proteins at concentrations lower than 500 fmol/µl is very inefficient. This is simply due to the fact that proteases exhibit 50% maximal activity in the 5- to 50-pmol/µl range. The alternative is to either develop chemical methods (but these still suffer from sample-handling losses) or perform the fragmentation directly in the mass spectrometer. MS/MS of intact proteins electrosprayed into various MS instruments has been achieved with a wide range of techniques, such as collisionally induced dissociation (Senko et al. 1994), infrared multi-photon dissociation (IRMPD, Little et al. 1994), UV laser-induced fragmentation and surface-induced fragmentation (Williams et al. 1990). All of these techniques can be performed using FT-ICR, where the high mass accuracy, resolution and the ability to perform MS^n experiments allows one to make sense of the complex fragmentation patterns. IRMPD allows a high degree of control over the energy added (or removed), is highly efficient and requires no gas loading. Thus, fragmentation can be controlled in a precise manner, which is important for MS^n experiments.

13.5.1.2 Protein Identification

It has already been demonstrated that intact proteins can be identified from their fragmentation in an FT-ICR by using a combination of the exact intact mass and a series of sequence tags extrapolated from MS^n experiments (Mortz et al. 1996). Recently, Li and Marshall demonstrated on-line identification of proteins by LC/ESI FT-ICR MS. A normal scan was first used to extract the exact mass of the intact protein and, on alternate scans IRMPD, was used to fragment one selected m/z ion from the protein. The intact mass was used with a wide mass window to select a subset of the database entries. The list of mostly b- and y-type fragment ions and any small sequence tags obtained from IRMPD were then matched against this set to identify the protein (Li and Marshall 1999).

13.5.2 An Alternative to 2D-PAGE? MS as the Second Dimension

2D-PAGE separates proteins according to their isoelectric points in the first dimension and by mass in the second dimension. Since stable isotopic-labelling methods have been introduced for quantification by MS, it is no longer necessary to run a second dimension sodium dodecyl sulphate PAGE gel for quantification by staining and scanning. It would seem logical, therefore, that one should replace the very low mass accuracy and resolution second-dimension gel with a high-resolution mass spectrometer. Initial steps in this direction have been carried out using direct scanning of first-dimension immobilised pH-gradient strips with an infrared laser (Ogorzalek Loo et al. 1997). One can even replace the first-dimension gel by capillary isoelectric focussing (CIEF), which can be directly connected to a mass spectrometer with an ESI interface (Jensen et al. 1999). FT-ICR allows high-accuracy mass measurement of the eluting proteins and quantification by isotope distribution; the proteins can be rapidly identified directly using the MS^n technique outlined above. Jensen et al. (1999) described the analysis of *Escherichia coli* cell extracts by CIEF-FT-ICR using injections of only 300 ng of protein, which is equivalent to three million bacteria (or 3000 human cells). Four hundred to one thousand putative proteins with a mass range between 2 kDa and 100 kDa were found. The sensitivity is approaching the range at which it will be possible to analyse individual cells. This can considerably reduce the amount of tissue it is necessary to obtain for sampling. For example, instead of analysing a whole-tissue biopsy, individual cell types can now be isolated by laser-capture micro-dissection to select only those cells exhibiting a morphology typical of cancer (Emmert-Buck et al. 1996).

13.5.3 Molecular-Interaction Mapping

The concept of the proteome, if restricted to the set of proteins being expressed in a cell at any given time, yields a fairly static picture. In reality, proteins can only exert their functions in a cell as a result of highly dynamic interactions with other proteins. The cell can be regarded as a series of interacting molecular machines that are formed from large protein complexes. The spatial and temporal modulation of these interactions is the key to defining cell functions in molecular terms. MS is now being explored as a tool with which to explore the dynamics of protein interactions. Monitoring changes in the phosphorylation states of proteins, especially those involved in signalling, can help define the set of interactions. A more direct method of defining protein–protein contacts is direct observation of the complexes and determination of the binding affinities. One method that has recently been developed is surface-enhanced laser desorption/ionisation (SELDI) affinity MS (Kuwata et al. 1998). The MALDI target is chemically modified to allow the attachment of a "bait" molecule (analogous to the bait protein used in the yeast two-hybrid system), which is then used to gather proteins that bind to the immobilised molecule. The surface can then be washed and the target placed in the mass spectrometer to analyse the materials that have bound to the immobilised molecule. A similar approach has been suggested for mapping the epitopes of antibodies by immobilising the antibodies on a target and presenting a digest of the target protein. The non-binding peptides are washed away, leaving the peptides that form the epitope.

An alternative method for determining protein interactions has been demonstrated by direct observation of complexes in a mass spectrometer. This has been termed bio-affinity MS (Bruce et al. 1995). This principle was used to screen for high-affinity ligands to the SH_2 domain of Src kinase (Lyubarskay et al. 1998). The protein was mixed with phosphorylated peptides, and the mixture was partially separated by capillary isoelectric focussing before analysis in an IT. Substances with masses higher than that of the target protein, indicative of complex formation, were chosen for MS/MS analysis. The ligand released by MS/MS was then isolated and characterised by further MS^n experiments. These MS-based methods are qualitatively useful for identifying ligands and can yield quantitative information. For example, the binding affinities of 19 peptides for vancomycin could be determined simultaneously using affinity capillary electrophoresis-MS (Dunayevskiy et al. 1998).

13.6 Conclusion

Hopefully, this book has given the reader an appreciation of MS in the realm of protein analysis and has helped point out the areas in which developments are most rapidly occurring. The development of programs like SEQUEST and the availability of mass spectrometers with a black-box software approach is enabling the application of MS by non-specialists. In 1989, the journal Science published an article entitled "Protein chemists gain a new analytical tool" (Barinaga 1989). The author quoted many highly respected mass spectrometrists, and some of the comments remain pertinent today. The bias was towards the most sophisticated MS instruments of that time, the tandem magnetic sector instruments, because "no-one but (Donald) Hunt has yet been able to extract sequence data from a quadrupole instrument". Ten years later, tandem magnetic-sector instruments are regarded as relics from a previous era, and QQQ, quadrupole-IT and Q-TOF instruments are the instruments of choice. When asked for his view of the future, Don Hunt was quoted as saying "protein chemistry laboratories will have two mass spectrometers for every sequenator". At least in the experience of this author, all three sequenators have been replaced by three mass spectrometers, and the throughput has increased not threefold but by three orders of magnitude. One comment made in 1989 by John Stults is especially pertinent now. "If you look at where we were even two years ago, there has been a tremendous increase in our capabilities. Who knows where we will be in two or three years from now?" This author certainly does not know. However, if one understands the fundamental principles underlying today's soon-outdated instruments, one will be able to keep pace with the rapidly ensuing developments.

If anyone asks this author's view of the future, or at least his wish, it would be that there will be an MS on every bench, not just for protein, nucleic acid or carbohydrate identification but for the determination of molecular binding constants, flux rates and diagnostic applications. He would also wish to have this in a package the size of a matchbox, attached to a computer with intelligent software so that the user does not realise there is a mass spectrometer.

References

Barinaga M (1989) Protein chemists gain a new analytical tool. Science 246:32–33

Blackstock WP, Weir MP (1999) Proteomics: quantitative and physical mapping of cellular proteins. Trends Biotechnol 17:121–127

Bruce JE, Anderson GA, Chen R, Cheng X, Gale DC, Hofstadler SA, Schwartz BL, Smith RD (1995) Bio-affinity characterization mass spectrometry. Rapid Commun Mass Spectr 9:644–50

Carroll JA, Penn SG, Fannin ST, Wu J, Cancilla MT, Green MK, Lebrilla CB (1996) A dual vacuum chamber fourier transform mass spectrometer with rapidly interchangeable LSIMS, MALDI, and ESI sources: initial results with LSIMS and MALDI. Anal Chem 68:1798–1804

Chen R, Cheng X, Mitchell DW, Hofstadler SS, Wu Q, Rockwood AL, Sherman MG, Smith RD (1995) Trapping, detection, and mass determination Coliphage T4 DNA ions of 10^8 by electrospray ionisation Fourier transform Ion cyclotron mass spectrometry. Anal Chem 67:1159–1163

Comisarow MB, Marshall AG (1974) Fourier transform ion cyclotron resonance mass spectroscopy. Chem Phys Lett 25:282–283

Dunayevskiy YM, Lyubarskaya YV, Chu YH, Vouros P, Karger BL (1998) Simultaneous measurement of nineteen binding constants of peptides to vancomycin using affinity capillary electrophoresis-mass spectrometry. J Med Chem 41:1201–1204

Emmert-Buck MR, Bonner RF, Smith PD, Chuaqui RF, Zhuang Z, Goldstein SR, Weiss RA, Liotta LA (1996) Laser capture microdissection. Science 274:998–1001

Feng B, Wunschel D, Masselon C, Pasa-Tolic L, Smith RD (1999) Retrieval of DNA by soft-landing for enzymatic manipulation after mass analysis in ESI-FTICR. The 47th ASMS Conference on Mass Spectrometry and Allied Topics, Dallas, Texas, 1999. CD-ROM. The American Society of Mass Spectrometry

Fromont-Racine M, Rain JC, Legrain P (1997) Toward a functional analysis of the yeast genome through exhaustive two-hybrid screens. Nat Genet 16:277–282

Goffeau A, Barrell BG, Bussey H, Davis RW, Dujon B, Feldmann H, Galibert F, Hoheisel JD, Jacq C, Johnston M, Louis EJ, Mewes HW, Murakami Y, Philippsen P, Tettelin H, Oliver SG (1996) Life with 6000 genes. Science 274:546–567

Guan S, Marshall AG (1997) Two-way conversation with a mass spectrometer: nondestructive interactive mass spectrometry. Anal Chem 69:1–4

Jensen PK, Pasa-Tolic, L, Anderson GA, Horner JA, Lipton MS, Bruce JE, Smith RD (1999) Probing proteomes using capillary isoelectric focusing-electrospray ionization Fourier transform ion cyclotron resonance mass spectrometry. Anal Chem 71:2076–2084

Kelleher NL, Senko MW, Siegel MM, McLafferty FW (1997) Unit resolution mass spectra of 112 kDa molecules with 3 Da accuracy. J Am Mass Spectr Soc 8:380–383

Kornienko O, Reilly PT, Whitten WB, Ramsey JM (1999) Micro ion trap mass spectrometry. Rapid Comm Mass Spectr 13:50–53

Kuwata H, Yip TT, Yip CL, Tomita M, Hutchens TW (1998) Bactericidal domain of lactoferrin: detection, quantitation, and characterization of lactoferricin in serum by SELDI affinity mass spectrometry. Biochem Biophys Res Commun 245:764–773

Li W, Hendrickson CL, Emmett MR, Marshall AG (1999) Identification of intact proteins in mixtures by alternated capillary liquid-chromatography electrospray ionization and LC ESI infrared multiphoton-dissociation Fourier-transform ion-cyclotron resonance mass spectrometry. Anal Chem 71:4397–4402

Little DP, Speir JP, Senko MW, O'Connor PB, McLafferty FW (1994) Infrared multiphoton dissociation of large multiply charged ions for sequencing. Anal Chem 66:2809–2815

Littlejohn DP, Ghaderi S (1986) Mass spectrometer and method. US Patent 4,581,533

Lyubarskaya YV, Carr SA, Dunnington D, Prichett WP, Fisher SM, Appelbaum ER, Jones CS, Karger BL (1998) Screening for high-affinity ligands to the Src SH2 domain using capillary isoelectric focusing-electrospray ionization ion trap mass spectrometry. Anal Chem 70:4761–4770

Marshall AG, Senko MW, Li W, Li M, Dillon S, Guan S, Logan TM (1997) Protein molecular mass to 1 Da by 13 C 15 N double depletion and FT-ICR mass spectrometry. J Am Chem Soc 119:433–434

Marshall AG, Hendrickson CL, Jackson GS (1998) Fourier transform ion cyclotron resonance mass spectrometry: a primer. Mass Spectr Rev 17:1–35

McLuckey SA, Van Berkel GJ, Goeringer DE, Glish GL (1994) Ion trap mass spectrometry. Using high-pressure ionization. Anal Chem 66:689AA–696AA and 737A–743A .

Morris HR, Paxton T, Dell A, Langhorne J, Bordoli RS, Hoyes J, Bateman RH (1996) High sensitivity collisionally-activated decomposition tandem mass spectrometry on a a novel quadrupole/orthogonal-acceleration time-of-flight mass spectrometer. Rapid Commun Mass Spectr 10:889–896

Mortz E, O'Connor PB, Roepstorff P, Kelleher NL, Wood TD, McLafferty FW, Mann M (1996) Sequence tag identification of intact proteins by matching tanden mass spectral data against sequence data bases. Proc Natl Acad Sci USA 93:8264–8267

Oliver SG, Winson MK, Kell DK, Baganz F (1998) Systematic functional analysis of the yeast genome. Trends Biotechnol 16:373–378

Ogorzalek Loo, RR, Mitchell C, Stevenson TI, Martin SA, Hines WM, Juhasz P, Patterson DH, Peltier JM, Loo JA, Andrews PC (1997) Sensitivity and mass accuracy for proteins analyzed directly from polyacrylamide gels: implications for proteome mapping. Electrophoresis 18:382–390

Senko MW, Speir JP, McLafferty FW (1994) Collisional activation of large multiply charged ions using Fourier transform mass spectrometry. Anal Chem 66:2801–2808

Senko MW, Canterbury JD, Guan S, Marshall AG (1996) A high-performance modular data system for Fourier transform ion cyclotron resonance mass spectrometry. Rapid Commun Mass Spectr 10:1839–1844

Senko MW, Hendrickson CL, Emmet MR, Shi S, Marshall AG (1997) External accumulation of ions for enhanced electrospray ionisation Fourier transform ion cyclotron resonance mass spectrometry. J Am Soc Mass Spectr 8:970–976

Solouki T, Emmett MR, Guan S, Marshall AG (1997) Detection, number, and sequence location of sulfur-containing amino acids and disulfide bridges in peptides by ultrahigh-resolution MALDI FTICR mass spectrometry. Anal Chem 69:1163–1168

Taylor S, Tunstall JJ, Syms RR, Tate T, Ahmad MM (1998) Initial results for a quadrupole mass spectrometer with a silicon micromachined mass filter. Electronic Lett 34:546–547

Valaskovic GA, Kelleher NL, McLafferty FW (1996) Attomole protein characterization by capillary electrophoresis-mass spectrometry. Science 273:1199–1202

Wei J, Buriak JM, Siuzdak G (1999) Desorption-ionization mass spectrometry on porous silicon. Nature 399:243–246

Williams ER, Henry KD, McLafferty F, Shabanowitz J, Hunt DF (1990) Surface-induced dissociation of peptide ions in FT-ICR-MS. J Am Soc Mass Spectr 1:413–416

Zhang Z, Guan S, Marshall AG (1997) Enhancement of the effective resolution of mass spectra of high mass biomolecules by maximum entropy based deconvolution to eliminate the isotopic natural abundance distribution. J Am Soc Mass Spectr 8:659–670

Subject Index

Automated MS/MS Interpretation 167
  algorithms 168
  brute-force 168
  graph-theory approach 169
  Lutefisk 169
  sub-sequencing 168
  validation of database matches 174
Automation
  HPLC-MS/MS 68
  SEQUEST 83
AUTOSEQUEST 132

Beowulf 138
Blast 168

Calnexin pathway 207
Capillary chromatography construction 60
  exponential gradient 63
  flow-splitting 62
  gradient delivery 62
  injection valve 63
  monolithic 62
  off-line loading 65
  polymer-based 62
  pre-formed gradients 62
  sample loading 63
  silica-based 62
Capillary Zone Electrophoresis (CZE) 76
  3-aminopropyl-silane 88
  electro-osmotic flow 87
  MAPTAC 88
  surface chemistry 87
CD48 glycosylation 225
CD59 glycosylation 222
Cell-map proteomics 259
Collision-induced dissociation
  in an ion-trap MS 15
  in a triple-quadrupole MS 15

Data controlled analysis 69
  dynamic exclusion 70
  instrument control procedures 93
  and peptide modifications 70
  and SPE-CZE 82

Database tasks
  integrated genomic analysis 252
  protein expression analysis 252
  protein modifications 253
  proteome mapping 252
Direct tissue analysis 34

Electrospray ionisation 16
  electrostatic model 56
  micro-electrospray 16
  nanospray 16
Enzymatic digestions, limitations of 111
Expression proteomics 259

FASTA 168
Fourier-transform ion-cyclotron resonance MS
  advantages of 262
  disadvantages of 264
  duty cycle 264
  high vacuum 264
  non-destructive measurement 263
  sensitivity and resolution 262
Future directions
  alternatives to 2D-PAGE 266
  molecular interaction mapping 266
  protein fragmentation 265
  protein identification 265

Gelatinase B glycosylation 226
Genome 1
Glycan structure analysis
  by enzyme arrays 216
  by normal-phase HPLC 215
  by reverse-phase HPLC 216
  by weak-anion exchange HPLC 214
Glycans
  chemical release of N- and O-glycans 209
  enzymatic release of N- and O-glycans 209
  in-gel release 212
  labelling the released pool 213

Glycosylation
   importance   207

Homology searching   176
   CIDentify   176

IgG glycosylation   221
Immobilised metal ion chromatography protocol   203
Interpretation of Spectra   143
Ion-fragmentation
   nomenclature   15
Ion-trap MS
   comparison with Triple-Quadrupole   28
   high-resolution scans   25
   instrument overview   24
   mass-range   25
   MSn   27
   product-ion scanning   26
   space-charge effects   25
Isotope-coded affinity tagging   6

Laser desorption ionisation   35
Lutefisk   169

Major histocompatability complex   34
Mass-analysis fundamentals   12
   ion-stability diagram   13
   Mathieu equation   12
   RF Field   12
Mass spectrometers
   advances in sensitivity   261
   miniaturisation   262
   types   260
   protein digestion   5
   quantification   5
   sample handling   4
   sensitivity   3
Matrix-assisted laser Desorption/Ionisation
   (MALDI)   35
   add-on surface   40
   chemical ionisation   36
   cluster decay   36
   contaminants   38
   conversion dynode   41
   delayed ion extraction   42
   dried-droplet   39
   flight-time focussing   42
   fragmentation   43
   high-energy collisions   43
   initial velocity   42
   instrument overview   40
   ion detector   41
   ion source   40

   ion-formation   36
   linear TOF   41
   mass analyser   40
   mass-resolution   43
   matrix   37
   orthogonal   42
   overlap model   37
   photo-ionisation   36
   picovials   108
   post-source decay (PSD)   43
   reflectron   41
   sample preparation   39
   velocity spread   42
Metabolome   1
Micro-ES emitter
   construction   57
   design   57
   high-voltage connection   58
   liquid junction   58
   metal film   58
Microprobe analysis   35
MS-Tag   167, 174

Neuropeptides   34

Peak parking
   CZE   92
   HPLC   67
PepFrag   174
Peptide Fragmentation
   a-ions   150
   aids to interpretation   159
   Automated MS/MS Interpretation   167
   charge-induced   147
   charge-remote   147
   cleavage by aspartic acid   149
   energy transfer   144
   high and low energy   144
   immonium ions   148
   internal cleavage   148
   isotopic labelling   160
   loss of C-terminal amino acid   150
   manual interpretation strategy   152
   mechanism of bond cleavage   146
   nomenclature   145
   side-chain losses   150
Peptide ionisation   143
Peptide Mass Fingerprinting (PMF)
   algorithms   105
   automation   107
   chemical modifications   115
   cross-species identification   113
   delayed ion extraction   107
   effectiveness of digesters   110

fuzzy-logic   108
hydrogen-deuterium exchange   115
iterative searching   115
mass accuracy   106
orthogonal data   112
post-translational modifications   109
removing modifications   110
strengths and limitations   106
PeptideSearch   162, 167
   tag construction   163
Phosphopeptide derivatisation   204
   negative ion mode   191
   neutral loss scan   191
   precursor ion scan   191
   selective MS scanning   190
   specific chemical modification   191
Phosphopeptide isolation
   immobilised metal ion chromatography
      189
Phosphorylation site determination
   beta elimination and MS/MS   197
   chemical derivatisation   198
   Edman degradation   188
   Enzymatic removal of phosphate   190
   MS/MS of native phosphopeptides   196
Post-source decay analysis (PSD)
   charge-tagging   50
   chemical derivatisation   50
   data evaluation and interpretation   47
   hydrogen-deuterium exchange   50
   In-source activation   44
   ion gate   46
   ion reflector   46
   ion-activation   44
   post-source activation   44
   sample activation   44
Protein and nucleotide sequence databases
   EST database   233
   GenBank   232
   OWL   233
   PIR   233
   SwissProt   230
   TrEMBL   232
Protein kinase   187
Protein phosphatase   187
Proteome   1
Proteomics database design
   agents   251
   browser   250
   communicators   251
   database interface   249
   editor   251
   MS data   248
   principles   245

protein identification data   249
sample and gel tracking   245
spot location and quantification   247

Resolving sequence ambiguities
   deuterium incorporation   178
   higher mass accuracy   178
   multiple MS (MSn)   179
   O18 labelling   179

Sequence determination ambiguities
   dipeptides isobaric with amino
      acids   173
   Isomeric and isobaric amino acids   172
   unusual and unknown fragmentations
      173
Sequence tagging
   chemical ladder sequencing   117
   enzymatic ladder sequencing   117
   ragged termini   116
SEQUEST   125, 174
   algorithm   128
   and post-translational modifica-
      tions   135
   candidate selection   128
   complex-mixture analysis   140
   contaminant filtering   136
   contaminant libraries   137
   cross-correlation   129
   DeltCn   131
   interpretation of scores   130
   output file   130
   spectral quality filtering   137
   spectrum pre-processing   128
   Subtractive analysis   140
   Xcorr   131
SEQUEST SUMMARY   132
SEQUEST-PTM   135
   differential modifications   135
   static modifications   135
Sherpa   164, 207
Sialyl Lewis-X   207
Software integration   71
   LabView   71
   virtual instruments   71
Solid phase extraction (SPE)   76
   extraction materials   84
SPE-CZE
   General design   76
   liquid junction   78
   optimisation   84
   pre-saturation   79
   protocols   96
   sample elution and separation   79

sample loading   79
sample overloading   81
transient isotachophoresis   81

Time of Flight (TOF)   35
Transcriptome   1
Triple-quadrupole MS
    instrument overview   17
    neutral-loss scanning   23
    post-translational modifications   21, 24
    precursor-ion scanning   21
    product-ion scanning   18
Two-Dimensional Gel Electrophoresis   2
    alternatives to   6
    proteome maps   2
    weaknesses   2
Two-dimensional Gel Electrophoresis data-
bases
    Argonne Protein Mapping   240

Danish 2D-PAGE   240
ECO2DBASE   243
Embryonic stem cells   243
Haemophilus   2D-PAGE   244
Heart 2D-PAGE   241
HSC 2D-PAGE   241
Human Colon-carcinoma   243
LSB   242
MitoDat   242
NIMH/NCI 2D-PAGE   241
SIENA 2D-PAGE   240
Swiss 2D   234
UCSF 2D-PAGE   242
Yeast 2D-PAGE   243
Yeast Proteome Database   242

Variable-flow chromatography   65
    advantages of   66

Printing: Mercedes-Druck, Berlin
Binding: Buchbinderei Lüderitz & Bauer, Berlin